Licensed to Practice

The Supreme Court Defines the American Medical Profession

JAMES C. MOHR

D0162073

The Johns Hopkins University Press

Baltimore

© 2013 The Johns Hopkins University Press
All rights reserved. Published 2013
Printed in the United States of America on acid-free paper
9 8 7 6 5 4 3 2 1

The Johns Hopkins University Press
2715 North Charles Street
Baltimore, Maryland 21218-4363
www.press.jhu.edu

Library of Congress Cataloging-in-Publication Data

Mohr, James C.
    Licensed to practice : the Supreme Court defines the American medical profession /
James C. Mohr.
        pages cm
    Includes bibliographical references and index.
    ISBN-13: 978-1-4214-1141-5 (hardcover : alk. paper)
    ISBN-13: 978-1-4214-1142-2 (pbk. : alk. paper)
    ISBN-10: 1-4214-1141-5 (hardcover : alk. paper)
    ISBN-10: 1-4214-1142-3 (pbk. : alk. paper)
    [etc.]
  1. Medicine—Practice—United States.   2. Physicians—Licenses—United States.
3. Medical laws and legislation—United States.   I. Title.
    R728.M637 2013
    610.69—dc23   2013010484

A catalog record for this book is available from the British Library.

*Special discounts are available for bulk purchases of this book. For more information,
please contact Special Sales at 410-516-6936 or specialsales@press.jhu.edu.*

The Johns Hopkins University Press uses environmentally friendly book materials,
including recycled text paper that is composed of at least 30 percent post-consumer
waste, whenever possible.

LICENSED TO PRACTICE

For Elizabeth

CONTENTS

Introduction   1

Prologue   3

PART ONE: Background

1 Medical Regulation in the United States through the
   Civil War   9

PART TWO: The Medical Society of West Virginia

2 Dr. Reeves and the Founding   25

3 Building the "True Church"   36

4 Challenges from Within   49

PART THREE: The Board of Health

5 Securing Legislation   63

6 Exercising Power   80

7 The Dents Confront the Board   92

PART FOUR: The Courts

8 The West Virginia State Supreme Court   109

9 Conflict and Enforcement   125

10 The United States Supreme Court   139

PART FIVE: Implications

11   American Medical Practice after *Dent*   155

Epilogue   179

Acknowledgments   185
Notes   189
Index   213

Introduction

Americans in the twenty-first century accept without question the assumption that doctors must be licensed to practice medicine. But that, of course, has not always been the case, especially in the United States. Under the Constitution, medical license laws had to be passed one at a time on a state-by-state basis, and not until the 1870s did any states begin to experiment in a meaningful manner with various forms of medical regulation. Far from seeming self-evidently beneficial at that time, the nation's early license laws provoked intense disagreements, both among physicians themselves and within state legislatures. Some doctors and policy makers openly opposed medical licensing; many others fought exceedingly bitter battles over what the requirements for a medical license ought to be.

In the early 1880s, West Virginia became the first state to enact and effectively implement a genuinely restrictive medical license law. That statute stipulated specific educational standards necessary for obtaining a license and imposed criminal sanctions against anyone practicing medicine without a license—the two cardinal principles that continue to undergird modern medical license laws today. Doctors who disagreed with those stipulations challenged the West Virginia law in court, first at the local level, then at the state level, and finally before the Supreme Court of the United States. While some people regarded the new law as a progressive reform designed to advance the nation's medical care and protect the health of the people, others saw it as a violation of constitutional principle based on disingenuous premises. For the future of American physicians, the stakes could hardly have been higher.

This book tells the story of that first effective medical license law: who fought to pass it, who opposed it, how it got through the legislature, how it was implemented, why it was challenged, and why the Supreme Court's resolution of the intense disputes it provoked proved to be one of the most profoundly significant events in American medical history. Though seldom mentioned in either medical histories or legal histories—and virtually unknown to most people—the high court's 1889 decision in the case known as *Dent v. West Virginia* provided constitutional ratification for nothing less than the conscious formation of a new profession in the eyes of the law: that of licensed physicians.

Even had the *Dent* case not produced such an important Supreme Court decision, the story behind it would be well worth telling. It is a riveting tale full of powerful personalities, intense emotions, professional ambitions, political dynamics—even a murder. The story behind this case also offers a look in microcosm at the effort of elite physicians to reengineer their place in American society during a crucial period of unprecedented economic reorganization and corporate consolidation in the United States generally. Once the Supreme Court ratified the West Virginia model, physicians were free to fight for similar license laws elsewhere, which resulted in decades of strikingly similar—if usually less dramatic—confrontations and policy battles in other states.

Collectively, the laws that followed West Virginia's model eventually transformed the practice of medicine throughout the United States from an unregulated *occupation* into a governmentally sanctioned *profession*. That transformation rationalized a previously chaotic medical marketplace and placed American medical practice on a rigorously scientific foundation. But that great transformation—surely the most significant development in American medical history—was neither smooth nor somehow predetermined. It would eventually produce a unique medical structure that resulted from conscious policy decisions like those made first in West Virginia. And even though the nation's system of medical licensing emerged at the state level, it could not have emerged at all without the imprimatur of the United States Supreme Court.

Prologue

On Saturday morning, March 7, 1891, Dr. George Garrison encountered Dr. George Baird near the busiest intersection of downtown Wheeling, West Virginia. They were by far the two best-known physicians in the city. Baird, a popular and prominent Civil War veteran who had once been mayor of Wheeling, was one of the most professionally successful and politically influential physicians in West Virginia history. A decade earlier Baird had been instrumental in securing passage of a law that created the West Virginia Board of Health. Both Baird and Garrison had served terms as gubernatorial appointees on that new state board during the 1880s. A generation younger than the sixty-two-year-old Baird, thirty-nine-year-old Garrison also enjoyed a flourishing medical practice in the city and was well regarded by the general public. Garrison was in the middle of his second four-year term as Wheeling's popularly elected city health officer, a post that allowed him to campaign for cleaner water standards and voluntary vaccinations at public expense.

Following a brief exchange of harsh words that morning, Garrison pulled a pistol from under his coat and fired two shots at Baird, who had just turned to tether his horse only a few feet away. One bullet entered the older man's skull below his ear and emerged through his left eye; the other plowed through his chest, severing a major artery before embedding itself in Baird's shoulder blade. Amazingly, with blood pouring from his eye socket, Baird managed to stagger into the nearest store. Once inside, he quietly removed his gloves and informed the stunned shoppers that Dr. Garrison had shot him. The proprietor of the shop helped Baird lie down on a table, where the influential doctor was pronounced dead twenty minutes later. Outside, Garrison turned to a passerby and said, "It's done," and then calmly repocketed his revolver and walked in a measured manner around the block to the city police building. There he told the astonished officer on duty that he had just shot Dr. Baird and had come to surrender himself.

Dr. George Garrison demonstrating how he shot
Dr. George Baird. *Wheeling Daily Register*, May 21,
1891.

Garrison and Baird had once been close personal and professional
friends. Baird was already a well-established doctor in 1869 when Garrison arrived in Wheeling as an eager, young would-be physician. The
older man had taken the younger man under his tutelage as an apprentice, which was the dominant form of medical training at the time.
Many people regarded their relationship as surrogate father and son.
When Garrison married and had a son of his own, he named the boy in
honor of Baird, his generous senior mentor. Baird, in turn, paid for Garrison to attend Jefferson Medical College, the school from which Baird
himself had graduated, so the younger man could obtain a formal MD

degree after his initial training. In short, their bonds had once been as strong and as close as might reasonably be imagined.

Disagreements over medical licensing, however, subsequently drove the two doctors apart—for reasons that will become clear in the story that follows. Fortunately, such disagreements did not lead to widespread violence throughout the medical profession; the lethal outcome in this case was personal and idiosyncratic on both sides. The two doctors' intense animosities over the issue of licenses had been building for five years, and the citizens of Wheeling knew that Baird regularly greeted Garrison in public with taunts, threats, and vicious personal insults. That this incident took place at all, however, is strong evidence of the intense emotions and high stakes involved in the transition from open market medicine to licensed professional practice in the United States.

PART ONE

Background

CHAPTER ONE

Medical Regulation in the United States through the Civil War

The Tenth Amendment to the United States Constitution declared that powers not delegated to the national government were "reserved" for exercise by the several states, or put differently, areas of public policy not specifically addressed in the Constitution were considered to be the domain of state-level legislators, not the national Congress. State legislators, in turn, could delegate their authority to local jurisdictions if they wished to do so. With few exceptions through the middle of the twentieth century, policies involving medicine and public health remained among those reserved powers and hence were hammered out on a state-by-state basis. Even today, both physicians and hospitals are still licensed by the separate states, not by the national government, and different states maintain different criteria for doing so.

States began exercising their reserved powers over matters pertaining to health soon after the Constitution went into effect. Through the first two-thirds of the nineteenth century, civil authorities often intervened directly in the marketplace by ordering the removal of foul nuisances and the relocation of businesses thought to be unhealthy for their nearby neighbors. Among other things, state and local officials also enforced ad hoc quarantines, employed physicians to care for the indigent poor, imposed regulations on the burial of the dead, and provided smallpox vaccination material at public expense. Maintenance of asylums for the mentally ill was the largest single item in the budget of some states by 1850. In short, from the early days of the Republic, lawmakers in the separate states paid attention to public health, as they understood it, and they did not hesitate to take measures to preserve and protect it.[1]

In sharp contrast to their active exercise of power in some areas of public health, however, those same American state legislatures consistently refused throughout the first two-thirds of the nineteenth century to require people practicing medicine to obtain a license. Consequently, anyone in any state who wished to enter the healing business was free to do so, and free also to style themselves "Dr.," regardless of their credentials or lack of credentials. Moreover, they were free to base their medical practices on whatever theories they wished to adopt and free as well to administer any therapies they thought might work, provided of course that the patient agreed to them. While the unwillingness to curb that open market in medical care seems downright irresponsible to modern Americans, it was not. It was a reasonable response to the practical realities of health care in the United States during the nineteenth century.

At least through the outbreak of the Civil War in 1861, almost all Americans routinely treated themselves and their family members for medical situations they were familiar with. Most families had a stock of standard procedures and home remedies passed along by earlier generations or borrowed from the experience of friends and neighbors. Most people also supplemented those procedures and remedies with advice they read in home health manuals. Self-help medical books of many different kinds were widely available throughout the United States, and they sold briskly. Only for acute, persistent, or clearly threatening health problems did people typically consult a physician. When they did so, they faced bewildering choices.

Through 1861, relatively few physicians in the United States had much formal education, let alone MD degrees from well-regarded institutions. Instead, the vast majority of actively practicing doctors had been trained by serving apprenticeships of varying lengths under the guidance of established physicians. They typically augmented those apprentice experiences by reading whatever medical textbooks were available to them, though the lessons and admonitions in one textbook often conflicted with those in the next. Some physicians also attended a course or two of lectures at one of the nation's many small and completely unregulated medical schools, often without staying long enough to graduate. Both the quality and the content of those courses varied enormously. Many physicians practiced only part-time, while also pursuing another occupation that provided the bulk of their livelihood. Between one-half and two-thirds of their bills remained unpaid at any given time, and payments frequently came in kind rather than in cash. Every doctor had a repertoire of tales involving long travels under miserable conditions to attend nearly impossible cases for little or no money—and most of those tales were probably true.[2]

Between two-thirds and three-quarters of American physicians through 1861 practiced some approximation of what they understood to be the standard procedures and accepted therapies taught in the nation's longest-standing and most prestigious medical schools. Collectively, these essentially generic doctors were known as "Regulars," a label they had applied to themselves. Regulars regarded themselves as the modern inheritors and contemporary interpreters of classical medical theories and learned medical traditions that stretched back to the Greek Hippocrates and the Roman Galen. Regulars had also organized and incorporated the vast majority of the nation's state and local medical societies, where they had opportunities to exchange useful information, present papers, discuss case reports, seek advice, and publish articles in medical journals.

The foundation of Regular practice had long rested upon the ancient concept of balance. In each human body, according to classical texts, four principal "humors"—blood, black bile, yellow bile, and phlegm—were thought to exist in a delicate relationship to one another. Whenever any of the four was either abnormally augmented or abnormally depleted, for example, by getting overheated or sleeping in the cold, ill health could result. To restore health, Regular physicians first tried to rid the body of contaminants by the use of various "purges, pukes, and sweats," many of which were convulsively violent. A patient might be miserable in the short run but surely had the satisfaction of knowing that the doctor had administered "strong medicine." Once the body was thoroughly cleansed, physicians then tried to correct the perceived imbalances. Those corrective therapies were administered on an individual basis according to specific characteristics such as age, gender, race, region, and season; and they ranged from dietary supplements and alcohol stimulants to bleeding, which was thought to reduce fever by bringing temporarily excessive blood supplies back down to normal levels.

Known as "heroic" therapies, those long-standing treatments were legendarily distasteful to the general public, and patients certainly thought twice before submitting to such often painful and unpleasant procedures. During the first half of the nineteenth century, many Regulars themselves also began to view the old heroic therapies with increasing skepticism, and a few leading Regulars—including some of the nation's most highly regarded medical educators—openly denounced the most violent applications as counterproductive and barbarous.³ As a result, many ordinary Regulars began to turn away from the old heroic procedures, or at least to use them less frequently by the middle of the nineteenth century. Case books and hospital records from the middle decades of the nineteenth century confirmed the growing tendency of Regulars to administer less violent emetics than the ones their immediate predecessors had used, along with the complementary tendencies to prescribe fewer depleting agents and more restorative substances.⁴

Even as the Regulars began to turn away from heroic therapy, rapidly developing fields such as chemistry and physiology began to produce laboratory results that had far-reaching implications for the understanding of medical biology. Pathological anatomists began to discover how various organs functioned and how they failed. A rising interest in medical jurisprudence and forensic medicine forced physicians to clarify their understanding of such diverse subjects as conception, gestation, and poisoning.⁵ Collectively, those new discoveries began to undermine the Greek and Roman theories that had sustained mainstream Western medicine for millennia. This left many ordinary Regulars disillusioned or frustrated, without effective treatments to replace their traditional procedures, and without a firm theoretical basis for developing more effective alternatives. They found themselves facing an uncertain situation that came to be characterized as "therapeutic nihilism."

For many mainstream Regulars, the most promising way out of their midcentury dilemma lay in a forward-looking commitment to science rather than a backward-looking acceptance of what was thought to have worked for past generations. Exactly what midcentury American Regulars meant by *science* was unclear and varied greatly from physician to physician. But for most of them a scientific approach to medicine involved paying attention to what could be objectively demonstrated, as distinguished from what fit existing theory or inherited wisdom; adopting a systematic and rational approach to therapeutic care, as distinguished from making an endless sequence of ad hoc judgments in each particular case; and applying new laboratory findings to medical practice, as distinguished from simply repeating procedures learned as an apprentice. Regular medical practice, in the words of the leading historian of these developments, increasingly shifted away from the concept of the "natural" toward a concept of the "normal," and away from the "individualization" of every case separately toward the application of generalized treatments based upon the assumption that all patients shared a large degree of biological "universalism."[6] Above all else, science offered hope, a belief that medical progress was possible.

While many ordinary Regulars were beginning to look to scientific research for help in combating the frustrations they faced in everyday practice, a minority of the nation's physicians diverged even further from traditional Regular medicine and embraced philosophically different approaches to healing. Some of those new approaches gained enough followers to support formal organizations of their own. All of the institutionalized alternatives explicitly rejected one or more aspects of traditional Regular medicine and developed systematic theories and therapies based on widely varying assumptions about the causes and cures of disease. During the first half of the nineteenth century, those formal alternatives included—among others—the Thomsonians, who advocated an every-man-his-own-physician ideology and the administration of a few easily gathered herbs; the Botanics, who also championed herbal medicines but trained physicians to identify and administer a far wider variety of them; and the Hydropaths, who claimed to restore health by administering water in various ways at various temperatures.[7]

Those three non-Regular medical sects were not trivial sideshows: Thomsonians at their height in the late 1830s claimed tens of thousands of local agents throughout the country. Botanics had established their own medical colleges by the 1840s and published their own national medical journals. Hydropaths not only published one of the best-selling popular magazines of the 1850s but also maintained an extensive string of water-cure spas throughout the nation, which were particularly popular with women. By 1861, however, two other groups had emerged as the most influential alternatives to Regular medicine: the Homoeopaths and the Eclectics.

Homoeopathy had been introduced into the United States from Germany in the late 1820s. Homoeopaths based their therapies on the theory that "like cures like"; hence they administered substances to sick people that would evoke symptoms similar to what they were experiencing in a person who was well. Hot peppers, for example, might be used against fever. The most doctrinaire of them further believed that those substances, when sufficiently agitated, could be administered in tiny—often absurdly minute—dilutions. While those substances were supposedly acting inside the body, patients were kept comfortable and given foods and beverages that restored strength. Adding to the appeal of the Homoeopaths was the fact that many of them were well educated in their own medical schools and typically knew as much about the human body as Regular MDs did. Many Americans, especially among the educated upper classes, found the benign approach of Homoeopathy far more appealing than the heroic therapy of the Regulars, and Homoeopathy boomed in the United States during the middle decades of the nineteenth century, particularly in the industrializing belt from the Mid-Atlantic states on across to the upper Midwest. In 1844, Homoeopaths in Philadelphia organized the nation's first national medical association, and in some states elected legislators pressed their state medical colleges to teach Homoeopathic medicine as well as Regular medicine.[8]

Eclectics, as their name implied, were open—with one important exception—to any and all therapies that seemed to produce results, whether or not they fitted known theories of disease. As the functional inheritors of earlier Botanic traditions, however, they adamantly opposed the administration of toxic chemical and mineral preparations, precisely the substances widely employed in Regular medical purging. Eclectics instead favored more gentle herbal potions and preparations, which—like the Homoeopaths—they combined with rest and restoratives. Some Eclectics were also reasonably well educated in their own separate medical schools, particularly in matters pertaining to medicinal plants. Eclectics could be found in formidable numbers all around the country but did not dominate practice in any one state.[9]

In addition to formally organized alternatives to traditional Regular medicine, a wide array of unorganized practitioners and quasi-physicians of many other sorts dotted the medical landscape of the mid-nineteenth century. Self-styled oculists treated eye ailments and made glasses; itinerant "cancer doctors" crisscrossed the countryside excising skin eruptions; countless individuals devised and sold potions and salves for almost any imaginable condition; and apothecaries prepared and prescribed their own medicines. "Indian doctors" claimed to possess indigenous Native secrets; faith healers promised transcendent cures; electromagnetic doctors extolled the healing properties of electricity; and neighborhood bonesetters treated fractures. In addition to attending births, regional midwives also provided medical care for women and infants. Column after column of the nation's daily newspapers advertised mail-order remedies and medical services, many of which were thinly veiled offers to deal with sexually transmitted diseases and reproductive problems. In short, notwithstanding the willingness of lawmakers to address other matters of public health, the practice of medicine in the United States remained a crowded, chaotic, and wide-open field through the beginning of the Civil War.

From the early decades of the Republic, the dominant Regulars, especially those who held formal MD degrees from mainstream medical schools, had repeatedly called upon state lawmakers to intervene in this increasingly diverse medical marketplace by requiring anyone who practiced medicine to obtain a license certifying their fitness to do so. They claimed that the ministrations of untrained practitioners were physically dangerous to the public, that charlatans were defrauding the ill-informed on many fronts, and that unscrupulous pseudodoctors were bilking the desperate by offering cures that could not work. In their view, the overall quality of medical care in the new United States would improve only when all physicians were held to what they regarded as their own high standards.

By 1830, lawmakers in several states responded to the Regulars by authorizing their state medical societies to issue licenses if they wished to do so. But even in those states, the same lawmakers refused to take the next key steps involved in any legally meaningful licensing system, the actions that would eventually lead to the *Dent* case in the 1880s: they refused to stipulate specific criteria for licensing and they were unwilling to deploy the criminal sanction of the state to punish people who practiced without one of those medical society licenses. The only advantage lawmakers in a few states were willing to grant physicians who held those society-based licenses was the right to sue for fees in state courts, a privilege they denied to unlicensed practitioners. But that was a hollow gesture, since suits for unpaid fees rarely succeeded.[10]

Thus, even before major rivals had arisen and coalesced into formal organizations, Regulars had been unsuccessful in establishing themselves as the exclusive agents of medical care. The licenses that some states had authorized them to hold were essentially honorific. That kind of imprimatur may have bestowed modest advantages in the marketplace—much as certification by a privately organized fraternity of organic gardeners might help sell vegetables at the local farmers' market today. But those privately awarded licenses cost money to obtain and did not prevent others from competing freely without them. Consequently, fewer and fewer physicians bothered to obtain licenses that were essentially window dressing in the first place. When rival groups began to organize formally in the 1830s and 1840s, they successfully pressured most of the state legislatures that had initially authorized Regulars to issue their own licenses into repealing even that symbolic concession.[11]

In 1847, leading Regulars moved to counter the rising influence of organized alternative physicians by establishing a national federation of their state and local medical societies, which they called the American Medical Association (AMA). Though the AMA would eventually become the national umbrella organization of Regular medicine writ large, it was initially founded for the primary purpose of upgrading medical schools. In the face of increased competition and their own waning faith in traditional measures, AMA Regulars came to believe that their best hope for simultaneously maintaining their dominance in the American medical marketplace and improving American health care lay in embracing science, not only as a promising path toward improved therapies, but also as the new foundation for medical education. Even though the vast majority of their practicing colleagues were apprentice trained and had not mastered specific scientific fields in a formal way, the AMA pushed from the outset for the teaching of formal courses in anatomy, chemistry, physiology, and related subjects in their medical schools, not just courses on how to diagnose ailments, what to do in various situations, and how to apply accepted therapies. That commitment to formal training in science would become the cardinal tenet of Regular medicine and the principal rallying point for Regular physicians after the Civil War.[12]

Scientific education, they hoped, would also provide a basis upon which to elevate the legal standing of physicians. Regular doctors, especially those holding formal MD degrees, surely considered themselves professionals in a general or cultural sense. And so, for that matter, did many of the physicians practicing Homoeopathic and Eclectic medicine. Most ordinary Americans, in turn, also seem to have regarded physicians—at least the well-established ones—as professionals in a loosely social or perceptual sense. Nineteenth-century Americans routinely and axiomatically referred to the "profession" of medicine and to physicians as "professionals." Elite doctors could thus claim to be professionals in a loosely rhetorical or cultural sense, and that status seems to have enhanced their standing in society. But even the most elite physicians could not claim to be professionals in the eyes of the law. With a few relatively unimportant exceptions, no American jurisdiction distinguished legally between the practice of medicine and any other wide-open occupation. Untrained country herbalists could set up a medical office in any town in the United States and compete for patients with the best-educated doctors.[13]

Indeed, through the 1870s, it can be argued that American law recognized only two occupational categories as professions: lawyers and military officers. Both had the legal right to determine for themselves who would be allowed to enter their occupation, so no one could plead a law case without the sanction of officially sworn courts, and commissioned officers were the only ones authorized to promote others. Both of those professions—precisely because they were legally recognized by the state—also possessed two other privileges that ordinary occupations did not. The first was the right to assess and police their own performance. Lawyers answered to judges for their professional behavior; military officers had their own system of courts-martial. The second was the right to be rewarded for their efforts per se, not for the result of their efforts. A lawyer, for example, was entitled to the same fee, whether the client won or lost; a military officer could be promoted for fighting skillfully, even though his side suffered a temporary defeat.

Some scholars might include clergy as a third legally recognized profession in nineteenth-century America, since clergy exercised some quasi-governmental authority (such as the right to perform marriages), and in some jurisdictions they enjoyed special privileges (such as exemption from militia duty). Moreover, clergy were certainly judged on their efforts, not the result of their efforts, since neither the damned nor the saved could return from the dead to attest to clerical effectiveness. But the nation's many separate private denominations were free to choose their own clergy by whatever criteria they wished, without state sanction, and importantly, persons recognized as clergy by one denomination were not permitted to deploy the criminal sanction of the state against others who wished to be recognized as clergy on alternative theological bases.

Physicians committed to the AMA agenda clearly recognized the difference between being professionals in a cultural sense, on the one hand, and having the powers that went with legal recognition, on the other; and they realized that the former ultimately meant little in the open marketplace of the American Republic. They knew they could not by themselves impose their standards upon everyone who wanted to practice medicine; they knew they could not by themselves make non-Regular medical practice a criminal offense; and they knew they could not by themselves absolve their colleagues from the standard obligations imposed upon other occupations, such as the responsibility for outcomes. To simultaneously transform the practice of medicine into a legally empowered profession and eliminate their intraprofessional rivals, leading Regulars realized they needed the state. Consequently, often working through the AMA or its constituent state medical societies, they renewed campaigns designed to persuade their state lawmakers that anyone practicing medicine should be required to hold a license. And, of course, they wanted those licenses to be based on criteria that would limit the medical marketplace to educated and science-oriented physicians like themselves. To their dismay, however, midcentury lawmakers continued to rebuff the licensing proposals they put forward.

Some Regulars blamed the lack of medical licensing in the United States on an irrational anti-monopoly mind-set prevalent among public policy makers, which had been blown out of proportion since the Jacksonian years. They castigated lawmakers for their inability—or cowardly unwillingness—to distinguish between efforts to advance the general welfare and attempts to acquire special privilege. Others blamed the lack of medical licensing on a pigheaded anti-elitism in the fiercely egalitarian American Republic, a sentiment they saw as shamefully exploited by political demagogues. Still others blamed the continuing lack of medical licensing in the United States on a pervasive anti-intellectualism in the nation's nineteenth-century ethos, which made Americans suspicious of formal education and a priori credentials, as distinguished from practical experience. A number of historians subsequently accepted many of those Regular contentions.

While anti-monopoly, anti-elitist, and anti-intellectual dynamics surely influenced American policy makers, the primary reason why state legislators remained unwilling to enact exclusive license laws was far more straightforward: lawmakers had no objective criteria upon which to justify such licenses. Though Regulars dominated the medical marketplace and championed the study of medical sciences, they could not demonstrate that either their approaches to healing or their superior knowledge of the medical sciences produced better patient outcomes, at least in the realm of internal medicine, than a host of alternative approaches, including many associated with out-and-out folkways that had no theoretically formal or scientific basis whatsoever. Wildly different theories, therapies, and remedies all seemed to work quite well for some patients at some times and to have little or no effect on other patients at other times. Plenty of self-taught and self-proclaimed doctors seemed to do as well for their patients as the most rigorously educated Regulars with the most prestigious MD degrees. Homoeopaths, in fact, amassed a better record treating epidemic diseases in the 1840s and 1850s than the Regulars.[14]

Nor, despite their repeated assertions, could the Regulars produce evidence that their principal rivals were wreaking harm upon the public. A few near-criminal mountebanks no doubt peddled dangerous concoctions, and rogue practitioners occasionally performed dangerous operations they knew little about. But those were isolated cases, not cases systematically associated with particular approaches to medicine; irresponsible mistakes were as likely to be made by Regulars as by their rivals. Consequently, Regulars could not make compelling arguments in favor of exclusive licenses for themselves on the grounds that patient outcomes would be better for the general public; nor could they make compelling arguments in favor of exclusive licenses for themselves on the grounds that eliminating their rivals would make the practice of medicine significantly safer for the general public.

In retrospect, none of the diverse practitioners operating in the nation's wide-open medical marketplace through 1861 could do much for their patients beyond the provision of what would now be called palliative care. And for palliative care, patients were arguably as well off— perhaps even better off—in the hands of a Homoeopath, an Eclectic, or a neighbor as they were in the hands of a rigorously trained Regular who followed the classical theories taught in the nation's oldest and most established medical schools. The Homoeopaths, Eclectics, and neighbors typically administered diluted teas, herbal tinctures, or warm soups, while the Regulars typically continued to administer emetics, cathartics, and purges, even if they were using less violent agents than they once did and prescribing more restoratives. The teas, tinctures, and soups prescribed by non-Regulars and neighbors probably did little good, but they seldom did much harm either, and they usually allowed a patient to remain comfortable while the body healed itself.

The emetics, cathartics, and purges administered by the Regulars, in contrast, sometimes proved dangerously depleting, probably weakened the patient, and may well have retarded the process of self-recovery. Notwithstanding the Regulars' turn away from heroic practices and ancient theories toward science and education, the most widely prescribed medication they administered to Union troops during the Civil War was a mercury compound that had powerfully cathartic effects. Voting delegates at the AMA's national conventions continued to sustain their organization's formal endorsement of therapeutic bleeding right through 1881, even though most Regulars had by then abandoned its use. Given those realities, state lawmakers had no rational bases upon which to establish medical standards, and hence no rational bases upon which to justify exclusive license laws. During the 1850s, in fact, not only were Regular-sponsored licensing proposals regularly rejected, but such key states as Louisiana, Massachusetts, and Michigan actually revived the older trend of repealing even their honorific license laws.[15]

Consequently, by the outbreak of Civil War in 1861, no meaningful medical license laws existed in any state. While the respective Surgeons General in that conflict tried to establish formal examinations for the appointment of military surgeons (as physicians were called) to their regular national army units, the vast majority of the roughly fifteen thousand men who served as military surgeons during the Civil War were appointed by state-level political officers to attend their own states' regiments. Some of those state-level political officers also tried to examine physicians but had little basis for distinguishing among them. As one historian of Civil War medical care put it, American physicians at that time "simply did not form a unified community based on a shared training experience or body of medical knowledge." And for many regiments, virtually any type of care was regarded as better than none. As horrific casualty rates continued to rise and more soldiers were being lost to disease than to combat, almost any practitioner of any kind could attain a medical appointment, provided he was willing to risk his life with the troops. A few who did so had no serious medical training whatsoever, though they gained a great deal of experience during the war and many continued in active practice afterward. State lawmakers in the North had far more important things to worry about than the legal status of doctors, including the survival of the constitutional system itself; and state lawmakers in the South were trying to grope their way through the chaos of conducting a massive rebellion. Under such circumstances, no state legislature spent time or energy during the war debating whether they should require licenses in order to practice medicine on the home front.[16]

When the carnage of the Civil War ended in 1865, therefore, the practice of medicine in the United States remained the same wide-open and unregulated occupation it had been since the American Revolution. But the nation's leading Regulars had not abandoned their strong desire to upgrade the legal status of physicians. Once the Republic was restored, groups of Regulars in several states turned back in earnest to the project of trying to secure exclusive license laws. Encouraged more strongly than ever by the national AMA, Regular medical societies renewed their efforts to persuade their respective state lawmakers to limit the practice of medicine to physicians who shared their commitment to formal scientific education as the key to medical advancement. One of the most determined of those postwar Regular groups had the added advantage of operating in a uniquely promising arena: the brand new state of West Virginia, where political power was unusually concentrated and all public policies had to be worked out de novo. Yet even there they would not be able to realize their objectives as quickly as they hoped they might, and the ultimate ratification of their goals would require a decision of the United States Supreme Court.

PART TWO

The Medical Society of West Virginia

CHAPTER TWO

Dr. Reeves and the Founding

The state of West Virginia came into being during the Civil War after influential political and economic leaders in the northwestern counties of Virginia managed a successful secession from their state's Confederate government in Richmond. Those leaders saw their future tied to the industrializing Ohio River Valley rather than the slave-based economy of the Tidewater, and they had long resented what they considered to be their somewhat awkward and second-class position in the Old Dominion. Consequently, after Virginia joined the Confederacy in the spring of 1861, northwestern politicians organized a regional revolt, cast their lots with the Union armies, persuaded the Lincoln administration and the United States Congress to recognize their separate existence, and emerged in 1863 as the nation's newest state. West Virginia statehood was thus the product of a successful secession, paradoxically engineered in the midst of a bloodbath being fought to suppress secession.[1]

The business and political leaders who engineered the creation of West Virginia selected the Ohio River city of Wheeling as their new state capital. They optimistically regarded Wheeling as the emerging coequal of Pittsburgh and Cincinnati, and they hoped their new state capital, now liberated from the plantation politics of Richmond, might successfully compete for preeminence in the Ohio Valley. Iron, glass, and coal—especially coal—rivaled timber and agriculture as West Virginia's economic mainstays. Commerce in the new state's northern counties followed the tracks of the politically powerful Baltimore and Ohio Railroad (B&O). Across the southern counties commerce followed the tracks of the equally influential Chesapeake and Ohio Railroad (C&O). West Virginia's small-scale manufacturers and independent farmers resembled their counterparts throughout the nation, as did those engaged in professional and quasi-professional occupations, including the practice of medicine.

At the time of statehood, few physicians in West Virginia, especially those practicing in rural areas, had much formal medical education, let

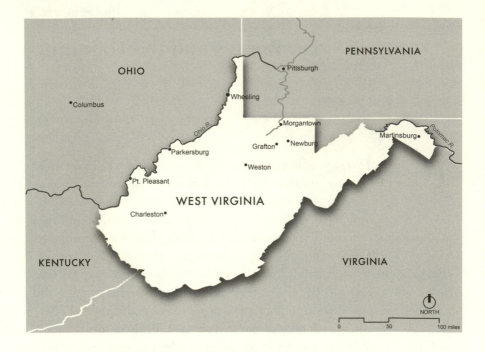

alone MD degrees. No medical schools existed within the borders of the new state, and only a small minority of West Virginia's physicians had previously attended medical schools elsewhere. Most of the people practicing medicine in the new state had the same rudimentary training as their prewar counterparts elsewhere: apprenticeships supplemented by medical textbooks, practical experience, and oral traditions. Outside the cities, most physicians practiced only part-time. If they were financially comfortable, their wealth almost certainly came from their collateral activities in commerce or agriculture, not solely from their medical practices.[2]

Especially in the rugged Appalachian valleys, family traditions also played a role in the medical field. Some extended families enjoyed a reputation for producing good doctors, in part because they knew the local customs, specific environments, and previous ailments of their patients and their patients' parents, all of which were regarded as significant factors in deciding how to treat each new case that arose. In a world of sketchy knowledge and uncertain outcomes, well-established multigenerational relationships built trust among rural families and helped assure patients that the doctor would try to place their welfare above pecuniary interests. When pressed, most rural physicians labeled themselves Regulars, though their actual practices and favorite therapies varied widely. Only a small minority of rural physicians openly identified with any of the formally organized alternatives to Regular medicine, and those who did so probably used the label as a way to gain attention, since almost none of them had actually studied at alternative medical schools.

In West Virginia's large towns and scattered cities, the practice of medicine was somewhat more formal than it was in the countryside, and more competitive as well. Physicians of all sorts tended to concentrate in urban areas because they could see more patients there than they could in the sparsely populated countryside or in isolated valleys. Moreover, in an era when the doctor generally went to the patient rather than the reverse, they could see patients expeditiously. Urban patients also tended to hold jobs that provided them with cash to actually pay their medical bills. Wheeling, a city of just under twenty thousand people in 1870, had twenty-six physicians in active full-time practice in the late 1860s. Three of them regarded themselves as Homoeopaths, three as Eclectics, and one as a Hydropath, while nineteen of them claimed to practice Regular medicine.[3] On a less formal basis, many other people in the city, including resident apothecaries and itinerant healers, practiced part-time and also proffered medical advice, various preparations, assorted drugs, specialized therapies, and practical assistance to any patients willing to pay for them.

Only the most elite physicians in Wheeling and other cities across the state held formal MD degrees from mainstream medical schools. Their degrees had been earned before the Civil War, principally at Regular medical schools in Philadelphia, Baltimore, and Richmond. Largely as a result of their similar training, these elite MDs looked askance at less formally trained physicians—including many of those who called themselves Regulars—and they vilified their overtly sectarian rivals as dangerous pretenders and quacks. Finding themselves practicing together in places like Wheeling, Fairmont, Clarksburg, and Triadelphia, the degree-holding Regulars recognized their kindred bonds and mutual interests. In clubs, civic organizations, and informal attachments, they nurtured their commitment to formal education, avowed their mutual faith in science, and expressed their distaste for much of what passed as medical care in the open marketplace. Even though the elite Regulars competed with each other for wealthy patients, they shared the same strong desire felt by their counterparts in other states to transform their unregulated occupation into a learned, science-based, and legally licensed profession.

To advance those goals, a small group of the new state's most prominent and best-educated Regulars decided in 1867 to organize a state medical society for West Virginia. Several local medical societies had survived the war in scattered locations around the state, and they continued to function. But the old Medical Society of Virginia (MSV) no longer reached into the new breakaway state of West Virginia. So the elite physicians recognized an opportunity to fill that void by creating a new state society of their own, one that would be capable of advocating the virtues of Regular medicine and scientific education more persuasively than any of the loosely structured and widely scattered local societies could hope to do separately—even if the latter were inclined to do so, which many were not. Additionally and importantly, because they were now operating in a distinctly smaller jurisdiction, these physicians also hoped that a new state medical society, based in the new capital city, might be able to influence health policy more effectively than the old MSV had ever managed to do in Richmond. Members of the latter, after all, had never succeeded in winning even the symbolic right to license themselves.[4]

The principal organizer of the initiative was James E. Reeves, a remarkable doctor who would quickly become the most influential physician in the state, and one of the most significant—if little-known—figures in American medical history. Though he was only thirty-eight years old in 1867 and a newcomer to Wheeling after serving in Fairmont during the Civil War, he had already established his credentials as a nearly fanatical proponent of Regular medicine and scientific medical education. Combining an unusually forceful and unbending personality with political acumen, tireless energy, and an almost messianic sense of himself, Reeves immediately assumed command of what would eventually become an all-out campaign to professionalize the practice of medicine in West Virginia. His was a name to be remembered by all who encountered him, and he would eventually engineer nothing short of a revolution in the legal status of physicians in his state.

Reeves had grown up in the town of Philippi in Barbour County, where his father was a tailor and a well-regarded Methodist minister of strict piety. Little is known about his childhood, but the devoutly resolute Reeves family seems to have instilled a fervor in their children that

Dr. James E. Reeves, founder and leader of the
MSWV. Women's Auxiliary to the West Virginia State
Medical Association, *Past Presidents of the West Virginia
State Medical Association, 1867–1942* (Charleston, WV,
1942).

produced a firm commitment to social causes, the ability to banish all
forms of self-doubt, a charismatic style of leadership, and expansive
organizational skills. James's younger sister, Ann Maria Reeves Jarvis,
organized nationwide "Mothers' Work Day" programs after the Civil
War, which were designed to promote an emphasis on pacifism and public
welfare all around the restored Union. Those programs, in turn, provided the institutional underpinning for her daughter's successful effort
to have Mother's Day proclaimed a national holiday.[5]

In 1843, when James Reeves was fourteen, his father stopped sending him to school and bound him out as a tailor's apprentice. The aus-
tere elder Reeves took that action because he believed every young man
should acquire a productive trade. But his son, who had been passionately fond of books from the time he was a little boy, found the work
mind-numbing. To broaden his horizons, James Reeves began to read

medical books in his spare time and quickly developed an intense intellectual interest in that field. Much as his Methodist forebears had experienced emotionally searing religious conversions, Reeves came to see medicine as his true calling, his holy vocation. So at the age of nineteen, against his father's wishes, James left the tailor shop and apprenticed himself instead to Elam Talbott, a local physician in Philippi, for one year, and then to Jacob Neff, a much more prominent physician in New Market, for another year. As was typical of medical apprenticeships in that era, he both assisted his mentors and attended some patients by himself. Neff, in particular, offered Reeves "every possible opportunity to study disease at the bedside, and frequently [gave him] charge of patients."[6]

In 1850, on the basis of what he had learned from his intensive reading and from those two apprenticeships, Reeves opened a practice of his own in the town of Sutton, which was on the main road from Morgantown to Charleston. There he encountered what he would have encountered in virtually every town in the United States at that time: intense competition from rival practitioners of several sorts. Young and unknown, Reeves attracted few paying patients. So a year later, he moved back to his home town of Philippi, where a construction crew was then building what became a famous covered bridge over the Tygart Valley River. When an epidemic of typhoid fever struck that crew shortly after Reeves arrived, he stepped forward to manage and contain the outbreak. His successful efforts gained him regional acclaim, thus assuring him a secure practice. In 1852 he went briefly to Richmond to take a term of medical courses at Hampden Sydney Medical College and then returned to western Virginia and continued to practice in Philippi until 1859. That pattern of reading, apprenticeship, practical experience, and limited exposure to formal instruction made Reeves's initial training similar to that of most physicians practicing throughout the United States at midcentury.

From the outset of his career, however, Reeves had been atypically earnest in his desire to play an active role in the advancement of medical science. While practicing in Philippi during the 1850s, he published well-researched articles in respected medical journals, and in 1859 he wrote a major treatise on typhoid fever, which was published in Philadelphia by the nation's leading medical printer. The book quickly became a standard text in the most prestigious East Coast medical schools. A classic autodidact, Reeves mastered medical Latin, became an expert in medical microscopy, and conducted what would now be regarded as experimental trials. Hungry for the latest scientific knowledge and for the imprimatur of a formal MD degree, he again traveled east in 1859 and enrolled at the University of Pennsylvania. There, in 1860, Reeves finally earned his coveted MD degree.[7] By the time he died in 1896, he had published twenty-two medical articles, four medical textbooks, and the nation's standard handbook on medical microscopy.[8]

Reeves's whole-hearted commitment to medical science combined with his training at the University of Pennsylvania to produce an equally whole-hearted commitment to the principles of Regular medicine, which he associated not with their classical past but with their scientific future. The vast majority of practicing Regulars throughout the United States in the late 1860s were content to follow general guidelines in their own practices and willing to overlook idiosyncrasies in the practice of others, but Reeves and the cadre of MDs he began gathering around him to form a state medical society abhorred even the slightest deviation from AMA guidelines. Since those guidelines forbade medical consultation with non-Regulars, Reeves not only refused to consult with non-Regulars about patients or cases but also refused to deal with any other Regulars who had anything whatsoever to do with non-Regulars, even on a friendly or a social basis. Since AMA guidelines stipulated the need for formal scientific training, Reeves was suspicious even of fellow Regulars whose experience was principally practical, and he was skeptical as well of MD degrees from medical schools that lacked a full catalogue of scientific courses, even if the school taught Regular medicine.

Looking back from the twenty-first century, such absolutism seems unnecessarily extreme, counterproductively arbitrary, and more than a little elitist. It also appeared then and now as blatantly self-serving, since Reeves and his associates were intent upon reshaping the entire medical profession in their own image and driving their rivals, both in and out of the Regular mainstream, from the field. In their defense, however, physicians like Reeves believed fervently that the Regular road toward scientific education was the only road to genuine medical progress in the long run. The Civil War had ensured the future of the Republic; science and education must now ensure the future health of its people. And even those who disliked Reeves's maniacal commitment to rigid Regularism and scientific education never doubted the sincerity of his desire to improve the nation's public health.

In 1868 Reeves persuaded the Wheeling City Council to establish a health department, and in 1869 he became the first permanent public health officer in the city's history. From that post, he worked tirelessly for the well-being of his fellow citizens, both rich and poor. He was especially alert to the growing health hazards of industrialization and became an outspoken champion of what would now be called occupational medicine long before that specialty existed. He eventually helped establish the American Public Health Association at the national level, and his successful campaigns to improve urban sanitation won him international recognition.[9] Samuel Gross, the famous Philadelphia surgeon, met Reeves at a dinner party many years after these events and characterized him as "enthusiastic, intelligent, and thoroughly in love with his profession."[10]

Reeves's messianic spirit energized and infused his organizational efforts in behalf of a new state medical society. Following preliminary contacts in the early months of 1867, he arranged for fifteen other prominent Regulars to sign a circular letter inviting "all members of the regular profession" to an organizational meeting. The letter exhorted physicians loyal to the principles of "noble science" to attend. All sixteen signers of Reeves's circular were from cities in the northern part of the new state, seven from Wheeling alone. Their intentions were clear in the first sentence of the circular. They were launching the society "as a means of elevating the standard of Practical Medicine and Surgery in West Virginia, and to render quackery odious, as it deserves."

Sounding remarkably like a modern sociologist, Reeves outlined ways a formal medical society might advance the group's ultimate objective, which he declared to be "our improved *status* [his emphasis]." A medical society, he argued, could foster learned exchanges at meetings and publish periodical literature, which would not only provide outlets for members already practicing scientific medicine but also function as vehicles for further educating and uplifting colleagues "of modest talent." Though the organizers were urban based at the outset, they anticipated informational campaigns in "our mountains and valleys" as well, because those areas were "strongholds of the enemy." Acting in concert, these Regulars hoped to persuade the people of their new state that a difference existed between themselves, "the intelligent physician[s]," on the one hand, and all the uneducated and non-Regular healers, "the murderous pretender[s]," on the other hand. If successful, Reeves triumphantly proclaimed, their proposed new medical society would improve "life and health" throughout the state and bring "respectability to themselves."[11]

The twenty-two physicians who answered the initial call convened April 10, 1867, in Wheeling. That date had been set because the organizers wanted to establish their new society in time to send a delegation to the AMA's annual convention, which was due to meet down the Ohio River in Cincinnati in May. Although the AMA was by no means the powerful force it would subsequently become in the twentieth century, it was already providing a national forum for people like Reeves, who were pushing the cause of formal education and Regular medicine at the state level. Sending a formal delegation to the national meeting would signal West Virginia's active interest in those efforts.

The social and political standing of the twenty-two West Virginia physicians who showed up for the organizational meeting in Wheeling could be judged by the fact that they were allowed to convene in the state capitol building and to hold their committee sessions in the state senate chamber. In order to show his "respect for the legitimate profession of Medicine and Surgery" and "to encourage the objects of the convention," the local head of the B&O gave the assembled doctors free passes to cover their travel to the meeting.[12] As it happened, many of them were already on retainer from the B&O, which maintained a corps of prominent doctors around the state to respond to accidents and medical emergencies along its lines. The doctors on retainer also provided expert opinions and sworn testimony when the railroads faced suits for injury.[13]

Reeves gave the welcoming address. At its core was a diatribe against all forms of non-Regular healing. He singled out Botanics, Eclectics, and Homoeopaths for special opprobrium and railed as well against the many charlatans who dared pose as Regulars but were not really qualified, in his opinion, to be one. Like many of the speeches that would be given before the Medical Society of West Virginia (MSWV) in the years to come, this one sounded as much like a fire-and-brimstone sermon as it did like a professional lecture. Medicine had one true path to salvation, according to Reeves, and that path followed science and education to the promised land of robust health, where doctors would serve as beneficent high priests, revered—and gratefully rewarded—by society. Regulars alone could guide the profession along that path. Anyone who wandered off it, even just a little, jeopardized both the future health of the American people and the future salvation of American medicine and therefore would have to be ruthlessly cast aside as a sinner.

True to his beliefs, Reeves openly challenged "the regularity" (his phrase) of an otherwise prominent and well-regarded Wheeling physician named John C. Hupp, who had come to the organizational meeting and wanted to join. Hupp practiced Regular medicine and was a member of the AMA, but his father-in-law was a physician known to favor herbal therapies over standard Regular treatments. Hupp was on friendly terms with his wife's father and hence technically in violation of the AMA proscription against any sort of consorting with non-Regular physicians.[14] For that reason, the purist Reeves thought Hupp should be rebuffed as a potential fifth columnist, but a majority of the other physicians at the meeting voted over Reeves's objections to extend Hupp an offer of membership. They were reluctant to see their nascent society embroiled in a nasty personal scrap before it could even get started, and besides, Hupp was acknowledged to be a kindly and socially progressive citizen, active in educational affairs as well as public health.[15] But the incident was a harbinger of things to come, and neither man forgot it. The assembled group then unanimously ratified a constitution that Reeves had drafted in strict conformity to AMA guidelines.[16] Thus, by the end of the day, April 10, 1867, the MSWV had come into being.

As soon as the new MSWV was formally established, the delegates elected Reeves state secretary, a post that would allow him to continue his organizing efforts. The honor of becoming first president of the MSWV went to William J. Bates of Wheeling, an ideal choice under the circumstances. Bates possessed credentials that would have stood out anywhere in the nation. The only son of a Quaker couple, he had grown up in York County, Virginia, and studied at the College of William and Mary as a teenager. Both his father and his grandfather had been physicians, and young Bates decided to follow in their footsteps. To do so, he had moved to Philadelphia to study medicine, and in 1835, at the age of twenty-three, he earned his MD degree with honors from the University of Pennsylvania, the same institution Reeves graduated from twenty-five years later. Bates had then traveled to Europe to continue his training in the most famous hospitals of London and Paris. While that pattern was not rare by the 1860s, it surely placed Bates among a small and elite subset of American doctors.[17] After practicing in southeastern Ohio, he had moved to Wheeling in 1842, organized the first city dispensary in 1845, and built one of the largest and most successful practices in the state. Now fifty-five years old, the dignified Quaker was a nearly perfect embodiment of the learned tradition of Regular medicine.[18]

Bates's inaugural address picked up where Reeves's welcoming address had left off. Bates too stressed the importance of limiting membership in the new society to avowed Regulars, and even then, they had to be strictly loyal to the principles of the AMA. He wanted only those who were "properly qualified to practice medicine; and whose professional character, professional standing, and professional rectitude [were] such as to make them honorable and worthy." Like Reeves, he believed in rigidly upholding doctrinal standards, and he strongly endorsed the group's creation of a Board of Censors both to police the purity of MSWV membership and to expel anyone caught consorting with the devils of sectarianism. After attacking the Homoeopaths, Hydropaths, and Eclectics in particular, he ended with a call for future legislative action that might make the nation's newest state a model for upgrading the practice of medicine throughout the country.[19]

Calls for legislative action, however, were premature. The newly established MSWV first had to solidify its own organization before it could hope to influence medical policy, even under the fluid political circumstances within their new state. That task would prove more difficult than Reeves and his associates had initially anticipated, owing in large part to their own doctrinal intransigence; and it would then be complicated by a national economic depression. And the Regulars of West Virginia in the late 1860s still faced all of the difficulties that their antebellum predecessors in other states had faced all along, including most significantly their inability to demonstrate that their educated and scientific approach to medical care was superior to the care provided by the practitioners they disdained.

CHAPTER THREE

Building the "True Church"

While many physicians in West Virginia regarded the Medical Society of West Virginia as a vehicle for sharing experiences and renewing friendships on an annual basis, James Reeves and the other hard-core Regulars who had organized the society in 1867 regarded it from the outset primarily as a vehicle for upgrading the practice of medicine in their new state. Instead of allowing anyone to try their hand at the healing business, the organizers of the MSWV wanted to restrict the practice of medicine to people they regarded as professionally worthy. Instead of allowing untrained individuals to call themselves doctors, the organizers of the MSWV wanted all physicians to have formal medical degrees from schools adhering to AMA guidelines. Instead of allowing doctors to prescribe whatever treatments they wished, the organizers of the MSWV wanted to make sure every physician adhered to what they regarded as the safe and scientifically valid principles of Regular medicine. Though the leaders of the MSWV pursued their goals with an essentially religious fervor, such radical transformations were not easy to engineer, even given the unique circumstances presented by the creation of a new state. Initial progress was slow, and they encountered problems within their own organization.

One of those problems resulted from lingering sectional tensions in their new state. The MSWV had been founded by physicians from West Virginia's northern counties, particularly the areas associated with the Ohio River Valley west of the Appalachian ridge, on the one hand, and the commercial arteries served by the B&O along the upper Potomac River and its tributaries east of the Appalachian ridge, on the other hand. The vast majority of the MSWV founders had been loyal to the Union during the recent Civil War. And almost all of them were closely allied with the politicians who engineered West Virginia's secession from Virginia and now controlled state politics from their capital at Wheeling. In the new state's southern counties, however, many of the most prominent and best-educated physicians had sided with the

Confederacy. The latter felt slighted by the new MSWV, even deliberately shunned, despite the fact that they too considered themselves staunch Regulars and equally committed to the principles of the national AMA.

To protest the self-asserted leadership of the Wheeling crowd, educated and elite doctors in the southern portion of the state began staging medical conventions of their own. Just as it looked like West Virginia might end up with two AMA-affiliated state medical societies working separately, the groups reached an accommodation. The MSWV publicly rejected any "distinctions founded upon the sympathies of its members during the late unpleasantness," officially declaimed the perception of regional exclusivity as an "unfortunate misunderstanding," and in 1870 formally welcomed the southern physicians into the MSWV with elaborate politeness. As a practical matter, however, during the next two decades the southerners seldom played a major role in the affairs of the MSWV, and only a few of them attended the annual meetings, most of which continued to be held in northern cities.[1]

A second and ultimately more serious problem facing those who wanted to use the MSWV as a vehicle for altering medical practices in their new state was the size of their organization, or rather its lack of size. By the end of 1867, the twenty-two founders had added forty additional charter members. In 1868 the society experienced a net gain of six more members, bringing the total to sixty-eight, and then added four more in 1869, bringing the total to seventy-two. Following reconciliation with the southerners, official membership jumped almost 40 percent to one hundred in 1870, though many of the southerners were members only on paper. Through the next fifteen years, however, formal membership in the MSWV never exceeded 115 physicians.

In comparison to the total number of physicians practicing in the state, those MSWV membership numbers were small. During the 1870s, various leaders of the MSWV thought that somewhere between 650 and 950 people were actively practicing medicine in the state, about two-thirds of whom claimed to be Regulars. Those estimates did not include itinerant healers, part-time amateurs, or the many individuals who did not practice general medicine but treated specific medical and quasi-medical conditions. By 1887, leaders of the MSWV placed the total number of medical practitioners in the state at 1,300. Thus, by their own estimates, well under 10 percent of the state's total practitioners, no more than 15 percent of the state's actively practicing physicians, and fewer than 25 percent of the state's self-declared Regulars belonged to the MSWV during its first two decades of existence.[2]

A third enduring problem resulted from the clash of strong personalities and ideological disagreements that surfaced regularly and repeatedly within the MSWV itself. The doctrinal fanaticism of James Reeves, in particular, provoked frequent confrontations. Even his warmest admirers acknowledged that he could be "quick, almost hot, in his resentments." In an otherwise laudatory speech, his eulogist noted that Reeves seemed to have "a natural fondness for controversy [that] sometimes led him to exhibit too much of the ardor of combat." The eulogist also conceded that Reeves exhibited a tendency "to judge harshly and uncharitably," and some of his "animosities . . . endur[ed] to the end."[3]

Reeves's effort to bar Hupp from the MSWV's organizational meeting was only the first of many similar actions over the next twenty years. Reeves recurrently placed names before the MSWV Board of Censors for investigation and possible expulsion. He targeted anyone alleged to have violated the AMA Code of Ethics and anyone who questioned any aspect of doctrinaire Regularism—along with anyone who questioned him personally. Because many others in the MSWV agreed with both Reeves's ultimate goals and his absolutism, the society's annual conventions could be dangerous events for the naive or the ill-informed. A Morgantown physician learned this the hard way when he presented a paper at the annual meeting in 1875. Partway through his paper, he was interrupted from the floor on a point of order. The interrupter insisted that the presentation be halted in midparagraph and urged that the paper be placed immediately before the Board of Censors as unacceptable material. The censors agreed with the challenger, the reading never resumed, and the Morgantown physician was dropped from the rolls.[4] Believing as they did that the future health of the restored Republic was at stake, Reeves and others like him spurned compromise.

Early presidents of the MSWV all took the same hard doctrinal line that Reeves insisted upon at the founding and that Bates, the first president, had heartily endorsed. In his 1869 presidential address, John Frissell boasted proudly that the MSWV had been "formed in an exclusive spirit." Had it ever. While Frissell wanted "every *regular* [his emphasis] within the state to unite" with the MSWV, he had no sympathy with several doctors who professed to be insulted after failing to receive membership invitations. Those doctors, in his view, had been deemed insufficiently Regular and hence unworthy of membership. "We want no drones, no grumblers, no ill-disposed persons, who will do nothing to build up the Society, and only belong to it to pick flaws, create dissentions, and destroy, if in their power, the work which others have done."[5] In 1870, Eugenius Augustus Hildreth's keynote address went so far as to suggest that the still tiny MSWV might nonetheless "need the application of the *pruning knife* [his emphasis] with an unsparing hand. . . . Only then [will] our noble profession grow into a 'goodly tree, and bring forth much fruit.'"[6]

Every presidential address for the next fifteen years continued to include similar rhetorical blasts against all forms of sectarian and irregular practice. The denunciations became an almost liturgical ritual, and leaders of the MSWV played a role similar to religious reformers exhorting the faithful. Some presidents performed the ritual using hyperbolic invective, while others used sarcasm. But none of them omitted that part of the annual service. Worth noting, however, was the fact that none of them ever used any actual data or specific evidence to substantiate the purported dangers they associated with alternative healers; none of them even offered anecdotal reports of particular patients being harmed by those they were hoping to drive from the state's medical marketplace. In that sense, their position was also nearly religious: their convictions were based primarily on faith. That does not mean they were insincere—far from it. As the medical historian John Harley Warner and others have suggested, these physicians saw in science a source of hope and optimism that might allow them to transcend the inadequate and frustrating therapies they had available to them. But they knew that science-based advances lay in the future, not in the present.[7]

At the 1872 meeting of the MSWV in Wheeling, President James M. Lazzelle, another close associate of Reeves, advanced the annual liturgy a step further than his predecessors had previously gone. Lazzelle began with the usual denunciations but then expressed his dismay that "*quackery* [his emphasis] flourished more successfully . . . in the past four years" than ever before in the history of medicine. The MSWV's ritual condemnation of medical sinners, in other words, was having little effect by itself; indeed, their evils seemed to be spreading. Those deteriorating circumstances in the medical marketplace—not the self-interest of MSWV members, he assured his colleagues—required forceful action on the part of the MSWV. Physicians who possessed the true faith would be reprehensibly irresponsible if they stood by passively while a growing number of pretenders led the public down medically dangerous paths.

The ideal remedy in Lazzelle's view would be "a higher standard of popular education," because he thought well-informed people should be able to recognize that much of what non-Regulars passed off as medical treatment had no logical basis. That premise conveniently ignored the popularity of Homoeopathy among many well-educated Americans, and Lazzelle acknowledged that a massive educational conversion of the multitudes was beyond the missionary powers of the MSWV in any event. Fortunately, however, he continued, "We have a *more immediate* [his emphasis] but perhaps less efficient remedy" available to us. That remedy was "legislation," by which Lazzelle meant a state-sanctioned medical license law based upon AMA principles. To head off the dangerous rise of medical heretics, in other words, the MSWV should consider trying to establish the medical version of a state church.

Though Reeves and his closest allies had founded the MSWV to push for the implementation of AMA principles, Lazzelle realized that he was nonetheless broaching openly for the first time "a mooted question" and possibly "treading [on] doubtful ground." After all, "some eminent medical men" were known to be leery of "laws regulating the qualifications of physicians for practicing their profession—and even some of the distinguished men of our own State share the same doubts." But in his view, "the growing tendency everywhere . . . on the part of the profession" was in favor of licensing, and "I think we can scarcely doubt longer its propriety." Those comments almost certainly referred to physician registration campaigns then getting underway in a few other states, though no state legislature anywhere in the nation had yet been persuaded to pass a mandatory license law on any terms, much less to give exclusive control of its medical marketplace to Regulars.[8]

The principal rationale for license laws, claimed Lazzelle, was public protection. "This is not class legislation, as suggested by its opposers. It is not for the benefit of the few against the many, but the very opposite. It is not for the benefit of the medical profession, as is so often falsely alleged, *but it aims at direct protection of the people against a hurtful class of men* [his emphasis]." The legislature needed to save their state's citizens from all who "tamper with the health and lives of the afflicted" for "base and selfish purposes." Unqualified physicians were "criminal offenders against public policy, and should be . . . promptly dealt with" like any others who threatened the public well-being. Lazzelle ended his call for action on a cautious and conciliatory note: "I leave this subject with you, and commend it to your careful consideration. . . . [T]hese sentiments are not deliberate recommendations, but merely suggestions." Still, he was convinced that statutory regulations were "of vital interest to *the whole people* [his emphasis] and to the profession of the State."[9]

Lazzelle was right in 1872 to recognize that large numbers of physicians, even among the Regulars, harbored serious misgivings about involving state legislatures in the business of medicine. Those misgivings were real, and they divided physicians bitterly in many other states. Similar misgivings would soon become apparent in West Virginia as well. But the hard-core leaders of the MSWV became convinced over the next several years that they would have to deploy the power of the state if they hoped to regulate the practice of medicine along AMA lines. Their resolve was strengthened during that period by their growing frustration with the behavior of the general public.

In speech after speech through the mid-1870s, the principal officers of the MSWV repeatedly castigated their state's ordinary citizens. The latter were not bewildered folks trying to make difficult choices in an uncertain medical marketplace; they were ill informed, easily duped, discouragingly superstitious, and pliable tools in the hands of those who employed cheap rhetoric and razzle-dazzle methods. Over and over, members of the MSWV told one another that the vast majority of West Virginians were simply too gullible and too ignorant to appreciate the value of real science, or even to understand their own best interests. Members of the MSWV came to believe that the general public—for its own good—would have to be forced by law to embrace scientific medical education as the only acceptable basis for medical practice in the long run. The general public was incapable of doing so on its own.

Though the MSWV's numbers were small, its leading members were fully aware of their politically powerful connections, especially with the economic interests that dominated state politics. At the meeting in Wheeling where Lazzelle first openly proposed a legislative campaign, for example, the delegates again came and went for free, thanks to passes from the B&O; were fawned over by the mayor and the governor; were treated to a concert by the Wheeling Municipal Institute; were given several days of factory tours by the city's leading industrialists; and were happy to imbibe from the best "private wine cellars" of the capital's leading citizens.[10] And they continued to hold their sessions, as before, in the capitol building. With good reason, leaders of the MSWV believed those connections could be translated into legislative action if properly mobilized. Presidents after Lazzelle, with the active support of Reeves and others behind the scenes, soon moved aggressively in that direction.

In 1873, President Robert H. Cummins suggested the desirability of specific laws to protect the profession from "suits for mal-practice [sic]," a growing problem at that time; laws to strike out against "quackery"; and laws to help physicians collect their fees.[11] But he stopped short of calling for a license law that would stipulate specifically who could and could not legally practice medicine. Cummins recognized that some members of the MSWV were still leery about opening the Pandora's box of legal licensing, for fear that an untrustworthy legislature might make matters worse. If a law passed that made licenses easy to obtain, for example, many of the very practitioners the MSWV wished to exclude might end up with state sanction.

In 1874, however, incoming president Silas W. Hall finally and un-ambiguously endorsed the idea of out-and-out licensing. Only through the power of law, he argued, could AMA standards be imposed and the practice of medicine upgraded. As long as alternative forms of health care existed, in his opinion, West Virginians would continue to patronize them. Consequently, to save the state's citizens from their own misguided choices, lawmakers should no longer permit those alternatives to exist. In defense of his position, Hall cited lawyers. Professional regulations were not as revolutionary or un-American as some seemed to fear, since the state, also for the good of the public, already imposed standards and requirements on the practice of law. The state now needed to do the same for the practice of medicine. The future well-being of the public's health was too important to let anyone who wanted to call themselves a doctor continue to practice—or in his view to prey—upon their gullible neighbors.

With the next biennial session of the state legislature scheduled for 1875, Hall urged the MSWV to launch a systematic campaign in favor of medical licensing on Regular terms. Local committees should be formed, legislators lobbied, bills drafted, and allies lined up. He hoped members would persuade their local newspaper editors to endorse the virtues of medical licensing, and he urged MSWV members to write letters to medical journals that would help colleagues see the merits of state sanction.[12] Hall's own direct efforts to enlist lawmakers, however, met with a monumental lack of enthusiasm. One member of the state house of delegates brushed off Hall's entreaties with the assertion that "no one would get sick or die until his time comes," licensing or no licensing. To put it mildly, as a chastened Hall reported back to the MSWV the following year, persuading a majority of the state's lawmakers that West Virginia needed restrictive medical licensing was clearly going to be a tougher job than he anticipated.[13]

Hall's successor, Matthew Campbell, found Hall's experience frustrating and perplexing, especially since Hall also reported that he had received little active support even from other physicians. So Campbell, another strong proponent of restrictive license laws, devoted his presidential address to suggesting specific arguments, tactics, and rationales that his colleagues might use in future legislative efforts. "[C]harlatanry and quackery, such as Homeopathy, Eclecticism and all other 'pathys and isms' that curse our age and country, continue to have their disciples and supporters," he began. "That these sects have no standard text books on Anatomy, Pathology or Physiology; that they have not in fact contributed a single great work on any scientific subject, is a startling and significant fact." And that fact, in his view, provided Regulars with their most powerful argument. Lawmakers needed to be convinced that "the thoroughly educated modern physician" was the only one who truly stood in "the sunlight of Science."

But standing in the glow of science would not by itself achieve the society's goals. For one thing, as Hall had learned to his dismay when talking to legislators in 1875, superstition and resignation regarding medical care seemed "inherent in the intellectual and moral constitution of our race." For another, Campbell believed that medical science had already outrun the ability of ordinary people to understand it. If the public were capable of recognizing the inherent superiority of scientific medicine, Regulars would not be experiencing the deplorable "nonappreciation accorded to us by the people at large." But they were not. And finally, he argued, "difficulties and embarrassments . . . grow out of our form of government, which . . . grants equal privileges to the man of science, to the knave, and to the fool."

The MSWV, Campbell concluded, should continue its public opinion campaigns, since any popular shifts in favor of AMA standards would surely help. But the MSWV was unlikely ever to generate a public groundswell in favor of the licensing program its members envisioned. Consequently, efforts to circumvent popular opinion seemed both necessary and justifiable. In the end—despite limited public support and despite the difficulties inherent in democratic decision making—the MSWV should concentrate its efforts on persuading a majority of individual state lawmakers, not a majority of all citizens, that the future health of West Virginia required a commitment as soon as possible to the champions of scientific medicine. Lawmakers should be persuaded that mandatory licensing on Regular terms would fulfill their solemn responsibility to protect the public health and safety of the people. In the absence of rigorous license laws, the dangerous and unscrupulous charlatans operating in the medical marketplace would not only continue to prey upon the innocent, steal business from the competent, and inflict physical harm in the present; they would also continue to resist real medical progress in the future.[14] Campbell's position steadily gained ground inside the MSWV over the next several years, as speakers spent less time in ritual denunciations of quackery and more time on the potential virtues of legislative intervention, even if the intervention appeared to be arbitrary rather than a reflection of the popular will.

By the time Hall and Campbell urged their colleagues in the MSWV to get actively involved in state politics, West Virginia's political circumstances had settled into a new pattern that would last another two decades. Unionist Republican majorities had engineered statehood, secured a constitution, and guided West Virginia's early years, but voters quickly grew disillusioned with the tumult of Reconstruction and the frustrations of postwar economic stagnation. In 1870, a heavy majority shifted dramatically over to the resurgent Democrats, who not only elected John J. Jacob as their party's first governor but swept every other statewide office as well. In 1871 the Democratic legislature eliminated voting restrictions on former Confederates, and in 1872 a lopsidedly Democratic convention redrafted the state constitution. The result effectively ensured control over the state legislature for the foreseeable future by the business-friendly wing of the Democratic Party, which was closely allied with West Virginia's powerful railroad and mining corporations.[15]

Though Jacob fell out with some of his personal rivals inside the Democratic Party, he managed to get reelected governor in 1874. The governor knew that most of the leading members of the MSWV had also by then abandoned their old Unionist ties and now sided openly with the overwhelmingly dominant Democrats. In 1875, as his term was nearing an end, Jacob suggested the possibility of an exchange of public health information between his administration and the MSWV. President Campbell regarded the overture as an unprecedented opportunity to publicly cement a political alliance with the state's ruling party, something that would allow the MSWV—in his words—to "wield a power for good in our state."[16] He invited Jacob to address the 1876 annual meeting of the MSWV, which Jacob did in warm tones. It was the first time the doctors had shared the forum of their annual convention with a non-physician in anything other than a ceremonial capacity, and the MSWV delegates made certain that the governor was armed with all of the arguments they would like him to make when talking with other Democratic Party leaders behind the scenes.[17]

When the legislature assembled a few months later, however, MSWV leaders correctly perceived that the 1877 biennial session would be thoroughly preoccupied with the relentless contraction of the state's economy, as the nation's longest continuously sinking depression grew worse and worse. The steadily deteriorating economic situation would come to a head later that summer in a massive strike against the B&O. When the strike turned violent, pitched battles and widespread arson virtually destroyed the city of Martinsburg in Berkeley County and eventually threw the entire state of West Virginia into armed turmoil. The deadly chaos ended only when the president of the United States answered a plea for help from Jacob's successor, newly installed Democratic governor Henry M. Matthews, and ordered federal troops to restore order.[18] So instead of proposing a medical license law, which would almost certainly be denounced as a plea for special privileges in the face of such hard times—or simply ignored altogether—Reeves and his colleagues planted two tactical seeds designed to bear fruit in more temperate times.

First, they had the MSWV delegates endorse a resolution to "*urge* [their emphasis] upon our State authorities the necessity of the establishment of a State Board of Health." Eight other states, following the lead of Massachusetts in 1869, had already established such boards, and the Regular state medical societies in those states regarded their boards not only as public health agencies but also as potential vehicles of Regular influence.[19] That perception had been strengthened two years earlier, in 1875, when the AMA itself had formally endorsed the idea of state boards. The AMA president sent a circular letter that year to the governors of all states without such boards, imploring them to create one.[20] Though West Virginia's economic doldrums doomed any chance of funding a new state agency in 1877, the case for a state board of health had been formally introduced, and the MSWV had gone officially on record in favor of creating one. Reeves would return to this project with great effect four years later.

Second, the MSWV created a permanent standing committee on legislation to coordinate its future political and lobbying efforts. And significantly, George Baird, another prominent Wheeling physician, agreed to head the committee. Baird was probably the most politically experienced and well-connected doctor in the state. He had been among the leaders of the West Virginia secession movement and served as mayor of Wheeling during the Civil War, when the ability to consolidate political power was a life-and-death matter. An outspoken advocate of industrial development, he was the principal stockholder of the influential Wheeling Bridge and Terminal Company. Baird was a close personal friend and business partner of Camden Johnson, a local Democratic chieftain who would become a United States senator in 1881; and he was an old school chum of James Blaine, speaker of the United States House of Representatives, who would run for president on the Republican ticket in 1884. Baird was also one of the highest-ranking Masons in the state.[21]

In addition to his industrial and political ties, Baird also enjoyed widespread acclaim among Wheeling's ordinary citizens. He had one of the largest medical practices in the city and was admired for his well-known habit of forgetting to bill his poorest patients. He would later be among the founders of an investment bank for workingmen. Baird had earlier succeeded Reeves as Wheeling's official health officer and continued his predecessor's progressive public health policies, especially with regard to the need for clean water and waste disposal in the nation's growing cities. He eventually emerged as one of the country's leading authorities on urban sanitation, though even in that realm he kept his eye on multiple objectives. At an international convention of the American Public Health Association in Toronto, Canada, for example, he gave a major address on the health benefits of paving city streets. In it, he argued that the very best pavers for health and sanitary purposes were being manufactured by a firm back in Wheeling; he did not add that he and other investors co-owned the firm.[22]

Baird had been a stalwart member of the MSWV from its founding, and he had long played major though less visible roles in the society's affairs. He was the one, for example, who had always arranged for the MSWV to hold its Wheeling meetings in the state capitol building. He had also served frequently on the Board of Censors, where he con-

Dr. George Baird, influential ally of
Dr. James Reeves. Women's Auxiliary to
the West Virginia State Medical Associa-
tion, *Past Presidents of the West Virginia
State Medical Association, 1867–1942*
(Charleston, WV, 1942).

sistently refused to compromise AMA standards. In this regard, Baird
clearly shared Reeves's absolute commitment to doctrinaire Regular
medicine, which was fortunate, since both men had intense and unbending temperaments. If anything, Baird was even more fiercely committed than Reeves to the goal of requiring all physicians in the future
to hold formal MD degrees based on a two-year curriculum of AMA-
approved scientific courses—a commitment so absolute that it would
ultimately lead to his murder thirteen years later.

Under the circumstances of the late 1870s, Baird's willingness to take
charge of the MSWV's overtly political activities boosted the spirits of

the society's inner core. While Reeves remained the unquestioned leader and public champion of the MSWV, the new committee on legislation had in the person of Baird someone with demonstrated political skills, both personal and economic ties to influential leaders in both major parties, and plenty of experience organizing successful political operations. And just as they hoped, Baird quickly set about laying the groundwork for future action. In each of West Virginia's fifty-four counties, he identified at least one physician willing to act as a local spokesman for the MSWV's legislative agenda. Looking forward to the next biennial session of the state legislature in 1879, Baird instructed each of his agents to "use all the influence" they could to persuade their local political leaders and their legislative representatives that the state needed a medical reform law.[23]

Before Reeves and Baird could mobilize their legislative campaign, however, they found themselves forced to overcome a serious challenge within their own medical society. That challenge would offer the Regular physicians of West Virginia an alternative definition of "science" and an alternative vision of what the future of the medical profession might look like. The man who mounted that challenge questioned the fundamental ideology of the MSWV founders, and in doing so, he ignited a personal feud between himself and James Reeves that would eventually help put West Virginia's medical license law before the United States Supreme Court.

CHAPTER FOUR

Challenges from Within

Along with the rest of the state, the Medical Society of West Virginia continued to suffer as the national depression of the 1870s dragged on. In 1877, the year President Rutherford Hayes ordered troops into West Virginia to suppress the state's violent labor protests, MSWV membership hit a ten-year low. Unemployed patients could not pay their bills, so marginally committed physicians stopped paying dues to an organization that did not seem to be accomplishing much in any event. In an effort to regain former members who had been automatically dropped for nonpayment of dues, the MSWV convention voted that year to welcome back any physicians who promised to pay in arrears, without charging the late-fee penalties stipulated in their bylaws. Even so, MSWV membership represented a smaller percentage of West Virginia physicians in 1878 than at any time since its inception.

Eugenius Augustus Hildreth, who was elected president of the MSWV at that low point, attempted to reverse the trend of falling numbers by launching a statewide recruiting campaign. He estimated that West Virginia had 612 physicians in active practice, including "376 to 400 *regular* [his emphasis] physicians, the remaining 236 are eclectics, botanic or Thompsonians, herb doctors, cancer doctors and homeopathists." Surely, he hoped, many of those 376 to 400 Regulars who were not yet members of the MSWV could be recruited. To help current members familiarize nonmembers with the objectives of the society, Hildreth had five hundred copies of the MSWV constitution printed. In particular, he wanted MSWV members to persuade the many preexisting county medical associations around the state that a rigorous license law promulgated from Wheeling would help them all by eliminating their economic competitors—all those shady interlopers in the medical marketplace, who "rob us of our just desserts [*sic*]."[1] That initiative, however, quickly backfired.

Most of the county medical associations in West Virginia were loosely organized and rather informal affairs. Though nominally Regular, many local societies welcomed any physicians who cared enough to get involved, and a few local societies were openly suspicious of the MSWV's doctrinaire exclusivity. The MSWV was on record as favoring licenses based on scientific competence, but most ordinary physicians had only limited formal knowledge of such subjects as chemistry, physiology, or pathological anatomy; the MSWV was on record as favoring rigorous educational standards, but most ordinary physicians were apprentice trained, and few had taken more than a course or two at any medical college. Plenty of physicians who had practiced for decades under the Regular banner now felt threatened by the prospect of license laws that might well exclude them as insufficiently prepared Regulars, along with all the non-Regulars, some of whom were their personal friends. Ironically, therefore, the more effectively the members of the MSWV conveyed their vision to local physicians around the state, the more doubts they raised among the very groups they had hoped to enlist as allies.[2]

From the point of view of James Reeves and his allies, the MSWV reached a nadir at the 1879 annual meeting. That convention assembled in the eastern panhandle city of Martinsburg, largely to recognize its successful rebuilding after the conflagrations of the 1877 railroad strike had all but destroyed it. The choice of Martinsburg also represented a tip of the hat to the B&O corporation, which essentially owned and managed that city. The B&O, after all, had consistently supported the MSWV by offering free transportation to meetings and by financing elegant receptions for its members, many of whom continued to remain available to the railroad on retainers. But Martinsburg was across the state from the MSWV's base in Wheeling, and only eighteen members of the MSWV showed up for the 1879 annual meeting—fewer than attended the society's first annual meeting twelve years earlier. And one of them left after the first day. An acting president had to be appointed to preside over the sessions, since Hildreth's successor in that office had moved his practice to Chillicothe, Ohio, where he hoped to make a better living. Moreover, many of the members who did attend the Martinsburg meeting were from the nearby valleys carved by Potomac River tributaries and hence out of touch with the Ohio River communities that were home to the MSWV's founders.

Trouble began immediately as the long-smoldering animosities between James Reeves and John Hupp again burst into open flames. Reeves had opposed Hupp's membership at the MSWV's organizational meeting on the grounds that Hupp was too tolerant of non-Regulars, but a majority of the other founders had accepted Hupp in the interests of harmony and unity. From that time forward, Hupp had quietly resented the imperious Reeves as much as Reeves resented the apparently apostate Hupp. Their personal enmity had previously broken to the surface in 1871, when Hupp became MSWV treasurer, normally an innocuous post responsible for collecting membership dues and reporting the organization's tiny annual expenses. Reeves was still secretary at that time and busily continuing his efforts to build the society's constituency around the state. When Reeves charged some of his recruiting expenses to the MSWV budget, Hupp seized the opportunity to challenge their legitimacy at the annual meeting. Reeves responded with invective rather than answers, and ugliness escalated to the point where the Board of Censors threatened to expel their chief founder and crusading secretary unless he apologized to Hupp, which Reeves eventually did in a stiffly worded note.[3] But Reeves remained alert for any chance to get even.

Now, eight years later, Reeves got his chance. Hupp and his father-in-law had helped establish the Wheeling Medical Society in 1868, which they hoped might serve as a less formal alternative to the long-standing and arch-Regular Ohio County Medical Society. Reeves, Baird, and the other leading members of the MSWV from Wheeling all belonged to the latter. In 1878 Reeves learned that the Wheeling Medical Society had begun to admit and consult with physicians who were not Regulars, a flagrant violation of both AMA ethics and the MSWV constitution. Like some of their counterparts in other states, the Hupps had decided that the hard times and the future of the profession called for a more tolerant and unifying philosophy than the rigid exclusivity endorsed by the AMA. In their view, the top physicians of all persuasions would have to work together to uplift the field. Otherwise, medical societies would continue to spend their time fighting one another, or they might simply wither away, as the MSWV seemed to be doing. That attitude was anathema to Reeves and his allies, who would rather have no society than one stripped of their own high standards. Reeves moved immediately to put allegations of misconduct against Hupp before the Board of Censors, hoping finally to expel the man he had mistrusted from the outset as a fifth columnist.

Six of the seven physicians on the Board of Censors submitted a motion to avoid further confrontation by referring the charges back to the local societies in Wheeling to sort out for themselves. But the seventh member, W. H. Brock, siding firmly with arguments advanced by Reeves and Baird, insisted that sanctions should be brought against Hupp. Following acrimonious discussion, two of the censors abandoned their own majority recommendation and joined Brock, Reeves, and Baird to engineer defeat of the majority report by a vote of 7 in favor and 10 opposed. That had the effect of putting the matter back before the convention as a whole. Former president John Frissell joined Brock, Reeves, and Baird by insisting upon strict adherence to AMA proscriptions and threatening to resign from the state society unless Hupp was removed for consorting with non-Regulars.[4]

Hupp counterattacked by accusing Brock of secretly obtaining railroad passes for himself and of openly advertising proprietary cures, which was another violation of AMA ethics. Brock had apparently run such ads when he was first starting out as a young doctor, but he had converted to the strict Regular line decades ago. On the face-saving pretext of needing time to review AMA guidelines on cases like these, the conflicted convention eventually voted to table the disputes until the following year. Ominously for Reeves and Baird, however, the delegates also reelected Hupp as treasurer, and only then did they turn to the reading of their medical papers.[5]

One of those papers offered a direct challenge to the central tenets of Regular medicine and hence to the hard-line leadership of the MSWV. Entitled "Relative Merits of *Vis Medicatrix Naturae* and Physic," the paper was read by a new MSWV member, thirty-year-old Arthur Melville Dent, whose practice was located in Weston. Arthur came from one of West Virginia's oldest and most influential families. Dents had been prominent since the American Revolution in the Appalachian counties that would later become north-central and northeastern West Virginia. Several generations of male Dents had been prominent in farming, mining, commerce, and politics. Arthur's father had sat in the Wheeling Convention that declared West Virginia's independence during the Civil War. But the widespread family was probably best known for its physicians.

Arthur Dent's great-grandfather had practiced medicine in the areas around Morgantown and Grafton through the early decades of the nineteenth century, and his grandfather, Marmaduke Herbert Dent, became the leading physician in nearby Preston County. Two of Arthur's uncles, George Washington Dent and William Marmaduke Dent, were also actively practicing medicine in West Virginia at this time, George in Monongalia County, William in Preston County. Both of those uncles were charter members of the MSWV. A third uncle, Felix Dent, was also a physician, but he had moved to Florida. Rounding out the family's medical network back home was Uncle William's twenty-five-year-old son, Frank Mortimer Dent. In 1876, Arthur's cousin Frank, who apprenticed under his father, had been elevated to a full partnership in his father's Newburg practice, the fourth generation in a direct line of Doctor Dents.[6]

Like so many other physicians of that period, Arthur too had trained first as an apprentice. He then went to Columbus, Ohio, in 1874 to take a year of course work at Starling Medical College, the same medical school where his uncle William had earned an MD degree in 1852. Arthur, however, returned to West Virginia before completing the number of classes required for a formal MD degree. He self-consciously valued practical ability over formal certification in any event, and he felt—like many others of that period—that he had learned enough in his first round of classes to be a good doctor, especially given his previous experience as an apprentice. In apparent confirmation of his decision, the practice he established in Weston had been thriving over the five years that had elapsed since then. In addition to his medical interests, Arthur was also a self-proclaimed radical freethinker and a contributing member of the National Liberal League, which advocated the absolute separation of church and state.[7]

At the heart of Arthur Dent's paper was the claim that Regular physicians had long overestimated the benefits of the drugs they administered. "It is doubtful," he contended, "whether the lists of mortality would materially swell if the physician should ignore nearly all so-called curative drugs" and simply let nature take its course. Most ailments, he believed, were probably "self-limiting," so doctors should be finding ways to strengthen their patients and helping them endure their ordeals rather than intervening aggressively with dangerous and unnatural preparations. Though a number of nationally prominent Regulars had been making similar arguments for decades, the administration of prepared drugs remained the bedrock upon which Regular medicine rested. Given the restive circumstances of the 1879 annual meeting, Arthur Dent's paper represented an overtly challenging alternative to the medical doctrines endorsed with religious fervor by Reeves, Baird, and the other founders of the MSWV. Indeed, as the hapless physician from Morgantown had learned in 1875, such a paper could not have been presented without protest at any prior meeting of the society. But the largely local delegates at the thinly attended Martinsburg convention of 1879—with Reeves and Hupp at each other's throats and with most of the Wheeling Regulars back home on the other side of the state—accepted Arthur's presentation.

The final official action of the 1879 annual meeting was the election of next year's president. That too boded ill for Reeves, Baird, and their allies because a majority of the members still present on the final day turned to Arthur's uncle, William Dent, a man clearly outside the Wheeling orbit. Based in Newburg, the president-elect's practice extended through the rugged mining and timber valleys of northeastern West Virginia. People in his region tended to look toward the Eastern Seaboard rather than the Ohio River, since their economy was based on shipping coal and lumber on the B&O down to Baltimore for distribution. Newburg itself, the largest community in Preston County at that time, was a rapidly growing, multiethnic coal mining city and a major stop on the railroad, where engineers reprovisioned and braced themselves and their locomotives for the steep grades ahead.[8]

Though William Dent was a charter member of the MSWV and a strong supporter of AMA principles over the long run, his attitudes toward the practice of medicine were known to be far less doctrinaire than any of the twelve presidents who preceded him. Though he held a formal MD degree from Starling Medical College, a Regular medical school, William agreed with his nephew Arthur that much of what was taught in such schools needed to be reassessed. Though he believed as strongly as his predecessors in the power of science to transform health care, William had a fundamentally different vision than they did of how the MSWV might help achieve that scientific transformation. The delegates who elected William Dent knew his views, and they knew he would expound his tolerant and expansive philosophy when he gave his presidential address at the next annual meeting.[9]

The 1879 meeting had thus offered Reeves, Baird, and the old inner core of the MSWV the greatest challenges they had ever faced within their own organization. Their recruiting efforts had raised new tensions rather than new members. Hupp seemed to be deliberately flaunting the society's rules, but the doctors present in Martinsburg had not only refused to expel him, they reelected him treasurer. Upstart Arthur Dent was allowed to read a paper critical of Regular therapies, which in turn guaranteed its publication in the MSWV's own *Transactions*. And the incoming president was known to be preparing a case for major changes in the basic orientation of the organization. So Reeves and his allies realized they would have to work hard behind the scenes, or risk losing control over the MSWV at the next annual meeting.

Consequently, the long-established leaders of the MSWV spent the next eleven months refortifying their positions. Reeves proselytized tirelessly in behalf of his "true church," warning against any deviations from its strictest dogma. Baird got himself reappointed to the Board of Censors, where he began reasserting the letter of the law, as he and his closest colleagues had laid it down from the beginning. The two of them and their allies also settled in advance on a reliable young acolyte named William F. Van Kirk as their candidate for president to replace Dent. Van Kirk was the nephew of former MSWV president Matthew Campbell and a graduate of Jefferson Medical College. He had loyally voted with Reeves in the latter's unsuccessful efforts in Martinsburg to expel the Wheeling Medical Society for tolerating non-Regular members, and he could be counted on to proudly toe the hard Regular line that his uncle had taken and his senior colleagues were now preparing to reassert.

As the 1880 annual meeting approached, the traditional leaders of the MSWV rallied their troops on the barricades of hard-line Regularism and prepared to repel any resurgence of the doctrinal tolerance that had surfaced in 1879 on the other side of the state. Reeves and his allies were right to be ready, because president Dent opened the meeting with just the sort of observations and proposals they had feared. Dispensing entirely with liturgical paeans in praise of AMA standards and the ritualistic condemnation of non-Regulars, Dent went immediately to what he regarded as the MSWV's biggest problem. In his view, fewer and fewer physicians around the state were willing to put up with the society's "want of harmony and mutual support." As a result, the MSWV suffered from "small attendance and the general lack of interest manifested by the majority of our professional brethren." In a pointed admonition aimed directly at Reeves, Dent told the assembled doctors that continuation of the personal animosities so painfully and publicly evident last year "must inevitably result in the downfall of our institution." Instead of clawing at one another and trying to get individuals expelled, all members, counseled Dent, needed to recommit themselves to "union and concerted action."

To correct what he regarded as the MSWV's increasingly self-destructive course, Dent offered a two-point program. First, members should remember what brought them together in the first place: their shared commitment to the ideals of science. And in Dent's view, a truly scientific approach to medical care would be open-minded and inclusive rather than doctrinaire and exclusive. He urged members to put aside their theoretical presuppositions and agree instead "that all experiments on health and disease, by whomsoever made, deserve calm investigation," even if made by non-Regulars. "The essence and nature of disease," he believed, "must be examined from different standpoints," because "the systematic arrangement of data" from all available sources and from everyone honestly contributing useful information offered the greatest promise of genuine scientific advancement. Dent hoped that the MSWV might stop thinking of itself as a star chamber designed to exclude the unworthy and begin to think of itself as "a tribunal of professional opinion for adjudication" of medical merit, "an arena of friendly discussion," a site of thoughtful assessment, and a forum for "friendly criticism." In modern terms, Dent's first proposal called for a transformation of the MSWV into an open and inclusive organization designed to continuously reevaluate and redetermine what would now be called best practice. Moreover, he hoped that young doctors would lead the way.

President Dent's second proposal addressed the question of how to bring the MSWV's "influence . . . to bear directly on the members of the medical profession, and indirectly on the public at large." The society's outside enemies were right, he thought, when they likened the social power and potential influence of physicians to "the priesthood of former times." Instead of denying the analogy, or being defensive about it, Dent believed that "this power [should] be uniformly exercised for the advancement of our profession, and the general good of humanity." And in his view, the best strategy for mobilizing that power—and obtaining results in the legislature—lay in rallying all of the state's physicians around something they could all agree upon: their occupational self-interest.

"There is no profession whose services are more freely rendered, whose duties are more exacting, whose labors are more destructive of [our own] health and comfort, than that which we have chosen. And yet it is a deplorable fact that the physician is the worst paid individual in every community," he claimed. "While in all branches of business numerous individuals are found, who have achieved financial success; a rich doctor is a *rara avis*, the wonder of his fellow citizens." Those who followed the "noble calling" of medicine were entitled to a better living

Dr. William M. Dent, the MSWV president
whose philosophies challenged Reeves's.
Women's Auxiliary to the West Virginia State
Medical Association, *Past Presidents of the West
Virginia State Medical Association, 1867–1942*
(Charleston, WV, 1942).

than most of them could manage under current conditions. The profession of medicine was "so unremunerative" that "a majority" of the state's doctors were "forced" to supplement their income from other sources. Therefore, Dent proclaimed in remarkably forthright terms, "it is high time that physicians form themselves into a solid phalanx, to resist the danger which threatens them financially. It is meet and proper, that they organize themselves into a self protecting guild, which should present a solid front to the encroachments daily made upon them by an exacting and ungrateful public." With unanimity, energy, and a well-conceived political strategy, Dent concluded, "the medical profession of this State may materially advance not only their interest, but also accomplish something towards the instruction of legislators in their duties to suffering humanity."

While Dent's all but revolutionary address was politely received, Reeves and Baird moved swiftly and firmly to counter his most threatening proposal. They had no intention of redefining science as an open-minded willingness to assess whatever came their way and no intention of transforming the MSWV into a more tolerant and inclusive organization than the one they had in mind. Since this meeting was in the Ohio Valley city of Parkersburg, the contingent of delegates from up the river in Wheeling was twice what it had been in Martinsburg, and the number of delegates from the eastern Appalachian counties was correspondingly reduced. That clearly favored Reeves and Baird, and in a situation where every vote might count, Reeves gained another ally by persuading the convention to seat his friend Benjamin W. Allen without having to pay his back dues; Allen was another graduate of Jefferson Medical College and would become president of the MSWV three years later.

Operating from his post on the Board of Censors, Baird had already engineered passage of a resolution declaring Hupp guilty of "gross misconduct as a member" of the society and also guilty of "forgery," by which the Censors meant an effort to cover up possibly censurable activities. As soon as the business session got underway, Baird introduced his resolution with a pointedly ironic stipulation: unless Hupp offered a formal written apology, as Reeves had once been forced to do, Hupp would "be and is hereby expelled from membership in this Society." None of the delegates ventured to oppose the Board of Censors report. Hupp then defiantly refused to apologize, and his name never appeared again on the rolls of the MSWV.

With that long-simmering feud finally resolved in their favor, Reeves and Baird nominated Van Kirk as the next president. There was no mistaking where the nominee stood. Van Kirk had already expressed the outrage he felt when he saw Arthur Dent's "Relative Merits" paper appear in the society's *Transactions*. "Better a thousand times that we should only publish the business proceedings of the Society," he fumed, "than to send forth commonplace matter, to be ridiculed by those capable of judging the merits of scientific and literary performances." He realized that the publication resulted from an MSWV policy that mandated inclusion of all papers presented at the annual meetings, and he recognized that amending the policy would be "a delicate subject." But even if a majority wanted to maintain the current publication policy for other reasons, he thought that the MSWV should at least start including a strong disclaimer in the *Transactions* stating that "the Society does not commit itself to the opinions expressed by the writers."[10]

Van Kirk declared emphatically his belief that the strength of the MSWV grew directly from "the excellent rules that were adopted at its birth, and have since been attempted to be rigidly enforced." Those rules had to be kept "pure" if the society hoped to upgrade the practice of medicine. "Like contagious poisonous breath," he stated, in a thinly veiled rebuke to the Dents, "offences have come, and will continue to come, so long as jealousy, avarice and vanity find a genial lodgement in the human heart." Against those "who are ready to pander to the prejudices of the multitude," he railed on, sounding like an Old Testament prophet, "let the condemnation be unmeasured . . . and the dignity of the society preserved." Van Kirk's subsequent praise for Reeves bordered on sycophantism.[11]

Once firmly back in control of the MSWV, with an uncompromising minion in the presidency, Reeves and Baird deftly appropriated the second of Dent's two proposals: that the time had indeed come for physicians to stop hanging back defensively and instead to mobilize their priestly powers in the political arena. Dent was right to declare that the time had come for doctors to secure a "self-protecting guild" for themselves, not only to improve their own economic well-being—as the outgoing president had suggested—but also to transform the practice of medicine from an open occupation to a legally empowered profession and to upgrade the quality of medical care in West Virginia. But Reeves and Baird, of course, envisioned a smaller and more exclusive guild than Dent had in mind, one that would serve the principal purpose for which they had established the MSWV in the first place: the advancement of Regular medicine as they defined it. And the best way to accomplish that would be passage of a mandatory license law based on AMA criteria.

So before adjourning the 1880 annual meeting, Reeves and Baird laid the groundwork for a decisive and well-coordinated political offensive designed to influence the 1881 biennial session of the West Virginia legislature, which was due to meet seven months later. Reeves agreed to assume personal command of an intense lobbying effort in the capital city, which he launched almost immediately. Over the following months, both in private meetings and at frequent social receptions he held in his home, Reeves talked with every state officer he could find, as well as every economic and community leader he knew. He paid particular attention to currying the favor of corporate officers, especially those from the rail-roads, most of whom had been friendly toward the MSWV from its founding. Reeves grew especially close to James H. Ferguson, chief legal counsel of the C&O, who provided legal and political advice. Widely regarded as the most influential attorney in the state, Ferguson was almost certain to be elected to the coming legislature himself; he would prove to be an invaluable ally.

Baird, in turn, resumed command of the campaign to generate sup-port outside the capital city. He again appointed two doctors in every county to make the case for state intervention in the medical market-place. Armed with informational brochures paid for by the MSWV, Baird's local agents were instructed to try their best to persuade both their skeptical medical colleagues and their overwhelmingly indifferent local politicos that medical licenses were essential to the future well-being of the state. Progress in the countryside was slow and uneven, but Baird's contacts reported modest gains for their cause over the following months.

Perhaps most importantly, before adjourning the 1880 meeting, the reestablished leaders of the MSWV also decided to run physician candidates of their own for seats in the next legislature. Campaigning through the summer and fall with the help of Baird's agents, six of the doctors who agreed to run won their elections in November 1880: for-mer MSWV president Andrew R. Barbee of Point Pleasant was elected to a seat in the next session of the state senate, and five other MSWV physicians—Isaiah Bee of Princeton, B. F. Irons of Pickaway, John B. Crumrine of Pennsboro, David Q. Steere of Pleasants, and W. H. Wayt of Marshall—won seats in the coming house of delegates.[12] Like Baird who recruited them, they were all physicians with recognized influence, political clout, and strong ties to the leading economic and corporate leaders in their communities. Isaiah Bee, for example, in addition to being one of the best-known doctors in the southern part of the state, had also been a delegate to the West Virginia constitutional convention and was a major investor in coal mining and railroad operations near his home city.[13] This time, vowed Reeves and Baird, the MSWV would be ready to act effectively.

PART THREE

The Board of Health

CHAPTER FIVE

Securing Legislation

With the legislative session of January 1881 approaching, James Reeves began drafting a bill that would achieve his goal of requiring physicians to obtain a license on rigidly Regular AMA criteria. Though he took counsel from other key members of the Wheeling medical elite and from his attorney friend James Ferguson, Reeves did most of the work himself. His first decision proved to be strategically brilliant. Rather than advance a license law by itself, Reeves revived the idea that he and the other voting members of the Medical Society of West Virginia had placed before the legislature four years earlier: the desirability of establishing a state board of health to oversee and coordinate the often confused and inadequate health efforts of local jurisdictions. Then last on the list of powers the state would delegate to the new board under the terms of his bill, Reeves included the authority to regulate the practice of medicine. Consequently, the most visible, potentially expensive, and politically sensitive elements of the proposed legislation appeared to involve issues of sanitation and quarantine, not the imposition of mandatory medical licenses.[1]

Reeves's strategy of linking medical licensing to other health and safety issues was not new. Regulars in several states, for example, also with the proactive support of the AMA, had tried repeatedly during the late 1860s and throughout the 1870s to link the need for medical licensing to the practice of abortion. Even though most abortions performed through the midpoint of gestation were not illegal at that time, and even though Regulars themselves were almost certainly performing most of them, the AMA officially assailed the widespread practice as a morally unjustified and physically dangerous procedure, and one that they alleged to be largely the work of marginal practitioners. Newspapers reinforced that impression with grisly stories about deaths following botched abortions at the hands of unsavory operators. Though the absolute number of such deaths was low, the sensational publicity afforded to cases like the infamous 1871 "Trunk Murder" in New York City provided lurid examples of how an unregulated medical marketplace threatened the public. Seizing the initiative, Regulars had urged legislators in several states to pass laws that would simultaneously criminalize most abortions, on the one hand, and ensure compliance by requiring medical licenses, on the other.

Reeves had supported the AMA's national campaign against abortion in an 1871 publication that blamed the high incidence of abortion on "the miserable starvelings who have missed their calling, failed of success by honorable practice, and are now fattening from the butchery of unborn children."[2] But the MSWV as an organization had not tried to make abortion a prominent issue in West Virginia, and, perhaps more importantly, Reeves also knew that lawmakers in other states had typically eviscerated the licensing clauses that physicians had bundled with their proposed anti-abortion bills. As a result, the practice of abortion had been criminalized in several states since the end of the Civil War, but the legislation that outlawed the practice had not included license provisions of the sort that AMA hard-liners were after.[3] Consequently, Reeves had recognized from the outset that he would need a more effective stalking horse than elimination of the state's abortionists. So he turned instead to the need to improve sanitation and protect the state from epidemics.

The first several sections of Reeves's board of health bill authorized the proposed new agency to inspect local drainage and sewerage systems; to impose standards of water purity in urban areas; to investigate and order the elimination of likely sources of disease; to halt all forms of transportation, commerce, and communication in the face of epidemics; to inspect and even to destroy domestic herds that might spread animal diseases; and generally to promulgate any edicts the board deemed necessary in the name of public health. The new board would also have ultimate jurisdiction over the activities of county medical societies. Those were potentially enormous powers, and they could be exercised solely at the board's "discretion." Anyone violating edicts of the board would be subject to stiff fines and lengthy jail terms. The board's authority to regulate the practice of medicine appeared only toward the end of the bill, and in comparison with its more visible public powers, that authority seemed secondary.

That strategic approach had at least four tactical advantages. First, Reeves himself, the bill's most aggressive champion and its leading lobbyist, had a long and completely credible track record of genuine interest in hygiene, sanitation, and preventive medicine. Shortly after the war, he had steered a public health proposal through the Wheeling City Council and then served as the city's first public health officer. He had helped establish the American Public Health Association, a national organization in which he remained extremely active. He had published and lectured on issues of sanitation and hygiene. Thus, Reeves's campaign for such an agency did not appear to be—and almost certainly was not—disingenuous or cynical. Though his prime goal was licensing authority, he also cared sincerely about public health.

Second, by 1881 the number of states with state boards of health had risen to nineteen.[4] In 1879, the United States Congress itself had created a national board of health to work with those state boards.[5] In almost all cases, the primary duties of those boards involved not the regulation of physicians but the collection of systematic data, the coordination of local health efforts, and the protection of citizens from epidemic diseases. Though a few of those boards also had limited oversight of the medical profession in their states, none had the degree of licensing authority that the Reeves bill proposed for West Virginia's board. Yet most West Virginia legislators were unlikely to realize the full implications of what Reeves had in mind for their state. They saw only that they were being asked to create an agency similar to those already existing in neighboring states, a type of agency even the federal government was encouraging. At key junctures during later debate over Reeves's bill, citizens urged passage to protect the public health. "If we are left to the mercy of our town and county authorities to make and enforce sanitary rules, there will be no protection worthy of the name 'Protection,'" wrote one. "The men that constitute the councils of our small towns are not sufficiently versed in 'sanitary science' or the 'art of preventing disease' to enable them to make sanitary laws to the best interest of the public."[6] In the end, the MSWV's opponents would attack the strong licensing clauses of the proposed board of health bill, but they were forced to do so in a context that made them look like they were weakening high-minded attempts to protect their state against epidemic disease as effectively as other states did.

Third, since issues like sanitation and quarantine involved the power to spend tax money, the power to restrict commerce, and the power to destroy diseased herds, those sections of the proposed bill would draw the most controversial fire on the floor of the legislature. Reeves shrewdly put those public health powers in the first eight sections of his proposed bill. Typical lawmakers understood those issues and wrangled at length over how to handle them. In contrast, questions about how to judge professional standards were an order of magnitude less obvious to the businessmen and farmers who dominated the legislature, and the language that allowed the board to grant licenses was not even introduced until all of the board's other powers had been enumerated.

Finally, neighboring Pennsylvania was said to be experiencing outbreaks of smallpox, particularly around nearby Pittsburgh. Those rumors, which Reeves and his allies shamelessly exaggerated, put added pressure on the West Virginia legislature to create a state agency capable of helping local jurisdictions combat what might be an impending epidemic. Reeves knew that many Southern states had created their state boards of health during the late 1870s for the explicit purpose of battling the yellow fever epidemics of that decade.[7] Now the threat of a smallpox epidemic beginning to fester just up the Ohio River from Wheeling might help him persuade West Virginia lawmakers to do the same.

Reeves's strategy worked well, as both the legislature and the popular press consistently referred to his proposal as a board of health bill, not a medical licensing bill. But Reeves made unmistakably clear in a later address to the MSWV that the creation of a board of health had never been solely an end in itself; it was principally a means of obtaining the power to license physicians.[8] From the outset, he envisioned the creation of a medical marketplace where the only doctors allowed to practice would be Regular physicians who had earned formal medical degrees from scientifically rigorous medical colleges. And importantly, though no previous legislature in the country had yet been willing to do so, he wanted the state to punish anyone practicing medicine without those qualifications.

Since members of the proposed board of health would be the officials who would implement and wield all the powers that Reeves had in mind, he had to make certain that the board would be composed of people who were likely to share his objectives. As a constitutional and political reality, appointments to such a board would have to be made by the governor. But Reeves did not want the board watered down with civil engineers, public safety officers, or veterinarians, much less with political cronies or non-Regular practitioners, notwithstanding the fact that the proposed board would have great authority over such things as sanitation, water supplies, quarantine, and domestic animals. So under Reeves's proposal the governor would be bound by law to appoint only

Regular physicians to the new board—no non-Regulars and no non-physicians—and those Regular physicians had to be "graduates of respectable medical colleges," had to have been in continuous practice for at least twelve years, and had to "have distinguished themselves by devotion to the study of medicine and the allied sciences."

The bill proposed three ways in which a person could gain a "certificate"—the word "license" never appeared—from the new board to practice medicine in West Virginia. The first was proof of graduation from "a reputable medical college," which was defined as a college the board "recognized as such." In practical terms, that definition would leave the veteran Regulars who got appointed to the new board free to recognize only well-established Regular medical schools as reputable, while rejecting degrees from sectarian medical colleges or flimsy diploma mills. "Reputable" medical schools, in the eyes of Reeves and those around him, were ones that followed AMA guidelines by requiring a thorough knowledge of the basic sciences and demanding a rigorous series of courses taken over an extended period of time.

A second way to gain certification from the board under the terms of the proposed bill was to pass an examination. The exam would be set by two members of the state board in conjunction with the president of the county medical society where the exam was taking place. At least superficially that seemed a reasonable and fair route to certification for would-be physicians who had already learned medicine through traditional apprenticeships or could not afford to go out of state for a formal education. But two added wrinkles in the examination clause revealed what Reeves was really after.

One of those wrinkles stipulated that MSWV members were deemed to have passed the exam automatically; they simply had to apply in order to get a license. The apparent logic behind that clause was the assumption that MSWV standards were so high that its delegates would not confer membership on any physician incapable of passing such an exam in the first place. The other wrinkle specifically required examiners to test candidates on "anatomy, physiology, chemistry, materia medica, pathology, pathological anatomy, surgery, and obstetrics" at "a sufficiently strict" level to justify public practice. That list of required subjects looked remarkably like the ideal Regular curriculum, as outlined by the AMA. Moreover, the examiners were free to set the exam in any form they wished and free to determine for themselves whether any given candidate's answers merited passing. Since the board members would be well-established Regulars to begin with, and since the county medical societies fell under the board's purview, correct answers would almost certainly be answers consistent with Regular medical practices. Consequently, the entire exam provision was yet another way to license only well-trained Regulars and exclude anyone who practiced on the basis of acquired experience rather than scientific knowledge. Even newspapers enlisted as supporters of the bill recognized that those two provisions were ultimately designed "not to allow anyone to practice medicine except a regular graduate."[9]

The third route to certification under the terms of the Reeves bill was submission of an affidavit proving that the applicant had "been engaged in the continuous practice of medicine in this state for more than ten years at the date of passage of this act." This grandfather clause was an out-and-out sop to hundreds of physicians already practicing around the state. Few of them were formal graduates of well-established Regular medical schools, and few of them would be capable of passing a rigorous examination based on hard sciences. But Reeves knew the legislature would never pass a law that summarily eliminated the practices of all those doctors. Among other things, such a law would leave many communities without physician services of any kind. So he included this ten-year rule.

Reeves did not like the ten-year loophole, and he said so. But he considered it a price worth paying to buy the support—or at least the neutrality—of the state's longest-practicing doctors, both Regular and non-Regular. Without such a clause, most of them would almost certainly oppose the bill, so nothing would get passed. With such a clause, many of them could gain the automatic imprimatur of a state-sanctioned license without having to change anything they did or pass any tests, so they might support passage of the bill. The clause was also shrewdly consistent with William Dent's idea of unabashedly appealing to the self-interest of as many of the state's physicians as possible. Moreover, as Reeves openly if somewhat callously remarked, its damage would be temporary, since those certified under the ten-year rule would eventually die off or retire, and all licenses issued after the first year of the board's existence would be rigorously based on the new criteria.[10]

Finally, to put teeth into the new regulations, Section 15 of the bill spelled out penalties for practicing medicine without a license from the state board of health. This was a crucial provision—perhaps the most significant clause of them all—and it would be at the heart of the Supreme

Court case that eventually arose from this legislation. American legislators in the past had sometimes been willing to authorize medical licenses and certificates of various sorts, but they had not been willing to punish unlicensed medical practice as a serious criminal offense. Reeves was determined to change that and to impose criminal sanctions broadly. His bill defined "practice" to include anyone who treated or prescribed for the sick, including even "apothecaries and pharmacists." Anyone convicted of practicing without a license would be subject to fines between $50 and $500, or imprisonment between thirty days and one year, or some combination of both. And those penalties applied explicitly to "each and every offense." Thus, if an unlicensed physician treated or prescribed for five patients in a single day, that single day could potentially cost the offender up to $2,500 in fines and up to five years in jail. Those were harsh terms indeed; Reeves meant business. Also in the bill was a sentence creating the new felony of filing or attempting to file a false diploma or affidavit regarding medical qualifications.

To shepherd his bill through the legislative process, Reeves turned to the man who had helped him draft it, his friend James Ferguson. In addition to being chief counsel for the C&O, which dominated the southern counties of West Virginia as effectively as the B&O dominated the northern counties, Ferguson had a long record of successful political management. He had served in the Virginia legislature and in the 1850–1851 Virginia Constitutional Convention, but he ultimately sided with the breakaway Unionists during the Civil War. After the war, he had drafted his new state's first legal code, and he soon emerged as the top railroad and land-law attorney in West Virginia. In 1870, like many other statehood Unionists, he returned to his original home in the Democratic Party, where his corporate connections immediately made him a power broker. Now head of the so-called Kanawha Ring, an influential clique that controlled the Democratic Party, Ferguson had the ability to pull political wires—a power acknowledged and envied by friend and foe. His opponents dubbed him "the evil genius" of West Virginia politics. Now back in the house of delegates for the 1881 session, the veteran attorney was the most influential figure in the legislature. One political observer characterized him as "the big dog with the brass collar."[11]

Neither Reeves nor Ferguson left any personal records of their alliance, so the reasons behind what soon emerged as Ferguson's sustained

James Ferguson, the powerful corporate lawyer and political party boss who supported Reeves in the legislature. George W. Atkinson and Alvaro F. Gibbens, *Prominent Men of West Virginia* (Wheeling, 1890), 282.

support for Reeves's proposals cannot be known for certain, especially since the board of health bill was by no means Ferguson's main interest in returning to the house of delegates in 1881 after a long absence from elected office. Ferguson was there principally to block farmer-labor-inspired railroad regulations and to orchestrate the election of Johnson Camden—his own law partner—to the United States Senate. Johnson Camden was also a close associate and investment partner of the MSWV's principal political organizer, George Baird. Perhaps Ferguson wanted to make sure he could count on the votes of the six MSWV physicians sitting in the legislature, votes that he got on the issues that most interested him. While in Wheeling, Ferguson had also suffered severe health problems, and perhaps he was rewarding the elite physicians who had arranged for his care. Moreover, independent of party politics, Ferguson would definitely have agreed with Reeves that education, especially higher education, held the ultimate key to improving American society. Ferguson had long been, and would long remain, a sincere champion of colleges and universities, including colleges for black West Virginians.

More intriguing, however, was the already well-established commonality of interests between West Virginia's railroad corporations and the state's most prominent physicians. Ferguson surely knew that his corporate clients had close ties with elite physicians all around the state.[12] Former MSWV president Matthew Campbell, for example, was a full-time B&O physician, and the sitting MSWV president, William Van Kirk, had previously been one as well.[13] The B&O also kept many prominent physicians in key positions on retainer, men like John Ramsay, the most successful Regular in Clarksburg and president of the Harrison County Medical Society.[14] Similarly close associations existed between the C&O and elite physicians in West Virginia's southern counties, where Ferguson's power was based.[15] Delegates friendly to the state's two dominant railroads would later insist on amending an anti-rebate proposal with a clause demanding "that nothing in this act shall be construed to prohibit any railroad or corporation by its officers, agents or otherwise from giving or offering to give to any member of the State Board of Health, or of any other Health offices of this state or of any county, district, town or city, therein, a pass or ticket to travel over its railroad or railway, or any part thereof, free of charge, or at a less charge than the usual rate for other persons."[16]

As a sophisticated corporate lawyer, Ferguson recognized that consolidation at the top—both industrial and professional—was rapidly becoming the American way. If it made sense for the best-administered rail lines and the most efficient mass-producing steel companies to gain control over small spur roads and local smelters, it probably made sense for the nation's top physicians to bring order to the practice of medicine. If the nation's future depended upon a healthy population of workers and a better standard of public health, it probably made sense to let scientifically educated physicians decide how to achieve those goals. In any event, and for whatever mix of reasons, the state's most powerful and effective political and corporate manager clearly decided to back Reeves and his plans for the reorganization of the medical marketplace. Given the dynamics of West Virginia politics in 1881, Reeves could not have had a stronger ally than Ferguson. Just as a new managerial class was engineering the corporate consolidation of American industry at this time, after nearly a decade of depression and uncertainty, Ferguson and Reeves were now cooperating to restructure American medicine. Theirs was an alliance perfectly consistent with the organizational dynamics of their era.[17]

Reeves arranged to have his draft bill introduced in the state senate by his friend and political ally Joseph Woods of Wheeling.[18] Reeves had also secured the confidence and support of the senate president, Albert Summers, another loyal member of Ferguson's Kanawha Ring. Former MSWV president Andrew Barbee, the newly elected senator from Point Pleasant's Mason County, was already working the senate floor. Summers referred Reeves's board of health bill to a friendly judiciary committee, whose members promptly called Reeves himself to testify in its behalf.

In a set speech that he subsequently released to the press, Reeves spent most of his time extolling the public health aspects of his bill. Nothing was more valuable than good health, he claimed, and the state could help people preserve that "invaluable estate" by creating a board of health, which could teach the people about sanitation and orchestrate attacks against disease. Under the guidance of well-qualified physicians, Reeves claimed, "at least three-quarters of the cases of sickness" routinely being experienced by West Virginians could be prevented in the future, and death rates would drop dramatically. Using estimates of wages then being lost to disease, Reeves argued that a board of health would not be a net cost to the state, but would instead be a major boost for West Virginia's economy, while simultaneously bringing a better quality of life to all classes. He also pointed out that many other states had already created boards of health, and West Virginia would look backward if it did not do so as well. He did not, however, point out the ways in which his proposed board would vary from all those other boards, most of which were civilian agencies and none of which had statutory authority to regulate the practice of medicine.

When he reached the medical practice clauses of his bill, Reeves rather disingenuously played down their implications. He made the board's regulatory powers look as innocuous as possible, intended to "strike at none but those wholly incompetent to assume and discharge the sacred trust of a physician." Just as the state restrained "the robber, the swindler, the assassin and the murderer," it should restrain "the ignorant pretender . . . who strikes at both life and purse—who can kill or cure, as luck may be." Echoing a refrain from several MSWV presidential addresses, he told the committee that these regulatory measures were necessary because "the people are less capable of judging of the qualification of medical men than they are of anything else brought to their notice, and, therefore, they need the protection which the passage of this bill will afford." To ensure high standards, lawmakers would surely want all appointees to the new board to be "graduates of respectable medical colleges and have had *twelve years* [his emphasis] continuous experience in practice." The time had come, Reeves exhorted, to "place West Virginia in the front rank of advancement" in matters of public health.[19]

Armed with Reeves's rhetoric, the judiciary committee promptly reported back to the full senate with a motion in favor of passing the board of health bill. Woods pressed for a second reading and offered a series of minor amendments. Most of them were housekeeping changes that clarified awkward phrasing in the original proposal, but a few were changes of substance. One of the latter required the board to make annual reports to the governor, rather than biennial reports in years when the legislature met, thereby increasing its visibility and potential influence. One of Woods's suggestions was rejected. That one would have further strengthened the hand of well-established physicians by increasing from twelve years to fifteen years the length of "continuous practice" necessary to be eligible for appointment to the board. By two votes, opponents of the board then scored another victory by persuading fiscal conservatives to join them in eliminating the board's state funding. Those skeptical of the whole project hoped no one would serve on such a board without compensation.

The judiciary committee incorporated the amendments and again reported the bill for its third and final reading. Debate began in earnest on Wednesday, February 9. Senator William McGrew from Morgantown moved to strike the word "regular" from the requirements for appointment to the board.[20] That motion carried by voice vote, a warning to Reeves against pushing too far. But the other stipulated qualifications, including graduation from a "reputable" medical school and a demonstrated commitment to medical science, remained intact. They would no doubt ensure the kind of appointments Reeves had in mind without stating specifically that all members of the new board had to practice Regular medicine. Several senators seemed uneasy, however, about the potentially enormous powers being granted to the new board, and they worried about how to pay for another state agency. Though

Barbee and Woods defended the proposal in vigorous speeches, inter-
mittent debate continued through Friday, when a cautious and somewhat confused majority voted to table the bill.

In response to that ominous development, friends of the bill worked
relentlessly over the weekend. They first persuaded the local newspapers to run editorials supporting passage. The *Wheeling Register*, for
example, endorsed the board of health bill that weekend as the golden
key to a "revolution" in public health. "We don't want to trust ourselves
in the hands of a blunderbus [*sic*] when disease is preying upon our
vitals, and it becomes a matter of life and death to have intelligent and
skillful treatment. At such a time we don't want any guess work or ex-
perimenting. We want a man to know his business when we put our
lives in his hands."[21] With financial backing from Wheeling's wealthiest
physicians, Reeves hosted the entire legislature at a grand reception
that weekend, which allowed the capital city's medical elite to button-
hole lawmakers one-on-one and make their arguments in behalf of the
board of health bill.[22] Plenty of private meetings took place as well.
Influential physicians from key districts, like E. L. Boggs of Charleston, took advantage of their free rail passes to go as quickly as they
could to the capital and join the lobbying campaign.[23]

Those weekend lobbying efforts and strong statements of support
from the press paid off on Monday morning, when the senate voted to
remove the bill from the table for reconsideration. Woods offered another ringing defense of the proposed board and then executed a successful parliamentary maneuver that put the bill permanently back
on the calendar. Over the next several days, Barbee directed discussion
of the measure and offered a series of minor amendments that clarified
procedural details. Senator Joel Stollings, a railroad investor and another stout supporter of the board bill, inserted a section barring itinerant healers from practice without board consent and fixed a stiff fine for
anyone who failed to comply.[24]

For many senators, the major sticking point remained financial. They
wondered how the new board could function without state funding,
they wrangled over how high of a fee physicians should be assessed
for certification or examination by the board, and they quarreled over
where such fees should go.[25] Senator William Dawson, who also represented Monongalia County, shared his friend McGrew's misgivings
about the idea of surrendering so much of the state's legal authority to
the Wheeling-based MSWV Regulars. He insisted on an amendment

that spelled out an appeal process for candidates who failed a board examination.[26] Then McGrew, the bill's most persistent and outspoken critic in the senate, persuaded his colleagues to strike the clause that automatically exempted members of the MSWV from that examination process.[27]

In the end, however, that sort of chipping away at the edges was all that skeptical senators could muster. In defense of the measure, Barbee offered a long speech that effectively stifled further discussion.[28] On Friday, February 18, the board of health bill passed the senate by a vote of 19 in favor to 0 opposed, with McGrew and four other senators pointedly abstaining. Barbee and Reeves had done their part in the upper chamber, which now sent the bill over to the house of delegates for concurrent approval.

Though Ferguson retained firm control over the lower house, the board of health bill faced considerable opposition from a vocal faction of anti-monopolist, anti-elitist, and anti-corporate insurgents, who constituted the farmer-labor faction of the Democratic Party—the faction that Ferguson had gone back into the legislature to thwart on issues of taxation and railroad regulation. Charles Ulrich, one of the elite Wheeling physicians who helped pay for the legislative receptions of the previous weekend and served as the MSWV's chief liaison to what remained of the Republican Party, characterized that farmer-labor faction of Democrats in the house of delegates as ill informed and misguided, the inadvertent champions of untrained "human leeches" practicing dangerous medicine. When public officials continued to condone such medical pretenders, he railed in a report to the MSWV, it was no wonder that so many people saw the entire medical profession as a "humbug" and preferred "to be 'tea'd' and 'poulticed' by some old woman" than risk treatment by a physician. Misguided or not, however, he also knew the bill's enemies in the house were not going down without a fight.[29]

Opponents of the board of health bill summarized their own arguments in a public letter posted on the front page of Wheeling's leading newspaper. The letter complained bitterly that the licensing provisions of the proposed bill were a thinly disguised effort on the part of MSWV Regulars to engineer a monopoly over the medical marketplace. That was abundantly clear, the letter pointed out, in the outrageous original clause that would have licensed MSWV members automatically. Moreover, the bill was cleverly worded to ensure that all board members would "come from one school—allopathy [Regulars]." An all-Regular "Board of Health, under this bill, can say that none but those presenting diplomas from the old or Allopathic [Regular] School are *reputable* [emphasis in the original] in their estimation," which would doom perfectly competent Homoeopaths and Eclectics. "Is that what is expected by the patrons of this bill?" asked the opponents, knowing full well that this was exactly what its patrons did have in mind. In a direct slap at Reeves, whom all parties acknowledged as the author of the bill, the letter angrily proclaimed that "whoever prepared this bill exhibits himself as possessing in a supreme degree the qualities of selfishness, arrogance and egotism."[30]

The protest letter was published anonymously and signed simply "Doctor." By itself, that would not have been unusual; newspapers routinely printed anonymous letters. Under the circumstances, however, the anonymity of the letter may have been more significant than any of the specific arguments it contained. In other states, such a letter would have been submitted under the imprimatur of a rival medical society and typically signed by the president of that society. But neither the Eclectics nor the Homoeopaths—the two leading opponents of Regular medicine nationwide—maintained a separate medical society in West Virginia. A handful of Eclectics had tried to assemble a state society of their own in the late 1860s—at the same time that Reeves was organizing the MSWV—but that Eclectic society was moribund by 1870.[31] Consequently, non-Regular medical opponents of the board bill had no institutionalized voice, no society to support them, and no established leaders to rally them or organize their protests.

The absence of organized medical opposition to the MSWV made the West Virginia situation unusual at this time, and it proved to be a major factor in the success that Reeves would eventually enjoy. Almost every other state in the country had either a Homoeopathic or an Eclectic state medical society, and several states had both. While those organizations were typically smaller and less powerful than the dominant Regular medical societies in their respective states, they could muster substantial political influence when they needed to do so. Several other states also had degree-granting Homoeopathic and Eclectic medical schools, which gave their physicians added clout.[32] As a result, Regulars had been unable to monopolize would-be licensing systems in any other state, even with the help of the national AMA. But the lack of organized non-Regular resistance in West Virginia significantly increased the chances that Reeves and the MSWV might be able to dictate their own terms.

Defense of the board bill on the floor of the house fell largely to the five newly elected MSWV physicians sitting in that branch of the legislature. Their only purpose for running in the first place had been passage of a medical license law; none of the five had ever been in the state legislature before, and only one of the five would ever run again. So now was their time to act, and they did so forcefully. As a local reporter put it, "The medical men in the House came up to the work nobly, and the passage of the bill proves the extent of their influence; their names will be revered by the medical profession in the state."[33] Reeves also publicly praised their "indefatigable, individual efforts."[34]

The behavior of John B. Crumrine of Ritchie County was illustrative. Prior to discussion of the board bill, the doctor had never spoken from the floor on any issue, preferring, in the words of his hometown newspaper, "to conduct himself in a dignified and discreet manner." When the board bill came up for consideration, however, Crumrine asked for recognition from the speaker and delivered an impassioned speech in its defense. He called it "the most important bill" of the entire session, stressed its public health aspects rather than its certification provisions, and called upon his fellow delegates to protect their loved ones from "the cold embrace of death" by passing "in the name of our proud little West Virginia" this "eternal monument to the wisdom and intelligence of this Legislature." According to newspaper reporters covering the legislature, this outburst of florid oratory from a previously sociable but essentially uninvolved delegate had a powerful effect on the house. The press noted that Crumrine had "been of great service [to] the medical profession," and the elite physicians of Wheeling feted him as a hero.[35]

Though they remained unhappy about the board bill, opponents had little chance to derail the Reeves express with Ferguson at the throttle. Only one delegate ventured an amendment. It would have eliminated the ten-year grandfather clause as grounds for certification, thereby provoking chaos among a majority of the state's practicing doctors. But that suggestion failed, and the senate bill remained unaltered. As a final effort to halt the bill before its third and final reading, J. L. Hall, a one-term representative from Barbour County who had opposed the whole board of health project from the beginning, tried to have the bill tabled. But that too failed.

In a well-orchestrated final tally, by a vote of 50 in favor to 10 opposed, the house of delegates then ratified the senate bill, creating a new state board of health. To the extent that the opposition fitted any patterns, one was geographical. Seven of the ten negative votes came from delegates clustered around Monongalia and Preston Counties, the areas long dominated by the Dent family and physicians like them. One of Preston County's own representatives, however, voted in favor, and the other followed the lead of his senate colleague and abstained. Ferguson, of course, along with all five MSWV physicians in the lower house, voted in the affirmative. Isaiah Bee, another of the MSWV physician-delegates, got the honor of informing the senate—a touching and symbolic gesture for Bee, whose own health was declining rapidly. The bill itself then went to the new governor of West Virginia, Jacob B. Jackson.

A cautiously conservative Democrat, Governor Jackson had been elected in November 1880, along with the six physicians in the current legislature. The son of a prominent attorney and a polished corporate lawyer himself, Jackson stood firmly with Ferguson's pro-corporate, pro-development wing of the Democratic Party and resembled the Southern politicians whom historians have often referred to as "Bourbon" Democrats. The new governor believed strongly that real progress in his state would require both a rational consolidation of economic resources under the leadership of knowledgeable elites and the training of skilled professionals.[36] He signed the board of health bill into law on the day it reached his desk, March 8, 1881, and promised to implement the act as quickly and effectively as he could. No one doubted that he would. The Regular physicians of Wheeling celebrated their victory at a grand party with mutual congratulations all around.

Newspapers in the capital believed they knew why the board of health bill had passed: "The success of the measure is due largely to the presence in the Legislature of representatives of the medical profession who were imbued with a high conception of its importance, and also to the co-operation of the outside fraternity, lead [sic] by Dr. REEVES [emphasis in the original], of this city, who has been unremitting in his efforts to secure the passage of the bill, and deserved the gratitude of the profession and the public for his services and sacrifices in this behalf."[37] George Baird saw things the same way when he summarized the legislative session for his colleagues inside the MSWV. The doctors on the floor of the house and senate were crucial in securing passage of the board bill, he acknowledged, and they deserved credit both for winning seats in the first place and then for securing policies that their organization had desired for a long time. But Baird agreed with the newspaper reporters that the heavy lifting had been done by Reeves, "the framer of the act." Reeves "watched its progress through both branches of the legislature with a steadfast interest; attending the sessions of that body daily to the neglect of his private business, and [was] always ready to counsel and advise with friendly members, and when necessary, to interview and win over opposing ones." He "contributed not only his time, but also sacrificed his pecuniary interests without the hope of fee or reward, to secure its passage."[38]

Thirty-six years after these events, Louis D. Wilson, who had been serving as the county physician for Wheeling's Ohio County when the original Board of Health Act passed, continued to reiterate the unanimous opinion of his contemporaries. In a private letter written to West Virginia's chief public health officer in 1917, Wilson, who was also the sitting president of the MSWV when the Board of Health Act went before the United States Supreme Court, attested that "the original [license] act was passed almost entirely through the efforts and influence of Dr. Reeves of this city [Wheeling]. . . . Of this I am personally cognizant."[39] The West Virginia Board of Health Act should probably have been named the Reeves Act.

CHAPTER SIX

Exercising Power

G overnor Jacob Jackson moved immediately to implement the Board of Health Act he had signed into law in March 1881. The law created a six-physician board—two each from the state's three congressional districts—who would serve staggered six-year terms. The law also required the governor to make those appointments within sixty days. Reaffirming his strong support for the project, Jackson announced his selections well in advance of the deadline.[1] The governor was known to be on friendly terms with Reeves and almost certainly conferred with him about whom to appoint. So no one was surprised when Jackson gave one of the full six-year terms to Reeves himself, who had written the law and coordinated the overall campaign to pass it, and the other six-year term to Andrew Barbee, who had led the fight for passage in the state senate.[2] Significantly, Jackson also filled all four of the shorter terms with staunch Regulars, exactly the sort of physicians Reeves had in mind when he drafted the statutory qualifications they were required to possess under the new law. As Reeves himself reported back to the Medical Society of West Virginia, "I am sure it is a source of pride to every member of this Society, and likewise to all regular physicians in the state, that the members of the State Board of Health, without an exception, belong to the 'True Church in Medicine.'"[3]

The six physicians appointed to the new Board of Health in the spring of 1881, however, almost immediately found themselves in an awkward position, facing two serious problems. First, the bill that passed the legislature had failed to allocate any state funds to their agency. Individually they were willing to serve without compensation from the state, and at a loss to themselves, since time spent on board business was income foregone from their private practices. But they wondered how the board's start-up expenses and administrative necessities would be covered. In theory, licensing and examination fees would eventually defray the board's ongoing efforts, but even that source of revenue was a doubtful prospect without at least some state funding to put new processes in place.

Second, the delegates and senators who composed the 1881 West Virginia legislature had voted to reassemble for an unusual "adjourned session" beginning early in January 1882. The principal purpose of the special session, which would have the same cast of lawmakers as in 1881, was a general overhaul of the state code. That compilation of statutes had not been systematically updated since Ferguson drafted its current iteration shortly after the war. Some of the provisions he had included in those tumultuous times now needed to be altered or dropped, and policies enacted since then had to be added. Because the brand new Board of Health law would be among the additions, its provisions could become the subject of renewed discussion and reconsideration.

Reeves had no intention of waiting to see what the adjourned session might do to his state Board of Health. More than most observers, he realized that the passage of his bill essentially intact had been, in his own words, "a happy surprise to the medical profession of the State." Moreover, he was acutely aware that the whole licensing project was, as he forthrightly acknowledged at that time, "an experiment."[4] He calculated that he and his colleagues on the newly appointed board had roughly six months to make that experiment produce favorable results, even without state funding, lest the upcoming special session have second thoughts about the "happy surprise" they had initially been persuaded to approve. Consequently, as designated secretary and head of the board, Reeves called the new agency officially into session on June 21, 1881, thus formally establishing its existence.

Reeves correctly perceived that many local newspapers had supported passage of his bill primarily because they wanted to have a state agency in place that might help protect their towns from epidemic disease. The editor of the *Point Pleasant Register*, for example, had hoped back in February, when the bill was pending, that "the medical gentlemen, aided by many sound thinking men in both Houses," would succeed in establishing "a State Board of Health, the better to be able to protect our people against epidemics of contagious and infectious diseases." The editor understood all too well that pestilence typically moved up and down the Ohio River, which fronted his city, yet the city officials had no control over what their counterparts in other riverfront communities did or did not do to help them contain spreading diseases. He added ominously, "We know not what the summer will bring forth."[5]

When the summer of 1881 brought forth renewed reports of smallpox in Pittsburgh, Reeves recognized the threat as an excellent opportunity to seize the initiative. If true, those reports placed cities like Point Pleas-ant, not to mention Wheeling itself, directly in harm's way, with the interior valleys not far behind. Consequently, with great public fanfare, the one-day-old board boldly declared a mandatory quarantine against any boats from Pennsylvania attempting to land anywhere along the Ohio River in West Virginia. Reeves and his colleagues also declared that people and goods crossing overland or by rail into West Virginia from Pennsylvania were subject to temporary detention and inspection. This preemptive action served as a dramatic announcement to the gen-eral public that the new state board did not exist merely on paper; its members were prepared to exercise their powers in defense of the public health, even if that meant an expensive and annoying interruption of vital economic commerce.[6]

The board also moved quickly to activate its regulatory powers over the practice of medicine. Reeves reminded his colleagues that they now had a legal duty to "guarantee to the people well-qualified physicians and surgeons." We must "soon," he warned, begin to "separate well-educated physicians from ignorant, dangerous pretenders, who cannot be otherwise regarded than as public enemies." Accordingly, board members also agreed at their first meeting to send formal announce-ments and registration forms to all known physicians in the state. The announcements explained the new requirement that everyone practicing medicine was now required to possess a license from the Board of Health. The forms they sent out invited proof of practice for ten years, the presentation of a medical degree, or an application for examination.

Along with the forms, the board sent a plea for voluntary contribu-tions to offset its own expenses until the legislature could be persuaded to authorize a permanent source of funding. Whether or not recipients took that request as a quid pro quo cannot be known, but the solicitation pro-duced more than enough money to cover the board's activities for the rest of the year. Reeves hailed those "contributions" as "proof of the perfect good will with which the passage of the law was received by those whom it was intended to control. Was there ever a like example," he asked rhe-torically, "of a State passing a law immediately affecting the privileges of an intelligent and influential class of citizens, who yet in response, and to show their pleasure, voluntarily contributed money from their own pockets to secure its strict enforcement?"[7]

To establish a precedent for the examination process, board mem-bers also conducted three examinations at their inaugural session in

June 1881. The three candidates were apparently handpicked in advance and standing by, well prepped for the symbolic purpose at hand. All three were apprentice-trained physicians who had been practicing as established Regulars for fewer than ten years but had not—or at least not yet—obtained formal MD degrees. One of the three was twenty-nine-year-old George I. Garrison, who had apprenticed under no less a figure than George Baird. Baird's young protégé was at that time regarded as an up-and-coming favorite of the elite Regulars in Wheeling, a perception that would change dramatically a few years later. All three initial candidates passed their exams, and on that basis, all three were granted formal certificates to practice.[8]

The following month, another apprentice-trained physician, Seymore M. Hopkins, presented himself for examination. Hopkins had been practicing in Wheeling but was not in favor with the elite Regulars of the city. Reeves and his board colleague from the First District, George Moffett, conducted the exam. This time, upholding their stated intention to keep standards high and to err only "on the side of safety to the people," they ruled that the aspirant had not passed. Taking advantage of the law's appeal process, Hopkins demanded reexamination before the entire board. But Hopkins must have seen the handwriting on the wall after Reeves and Moffett also rejected three of the next four applicants they examined. Instead of showing up for his second test, Hopkins decided to head down the Ohio River, where he became for the next forty-five years "one of the best known and highly esteemed physicians in Northern Kentucky."[9]

Reeves and his colleagues on the board remained remarkably busy through the rest of the summer and into the fall filing forms, certifying credentials, and administering examinations. By the time the adjourned session of the legislature reassembled in January 1882, the six-month-old Board of Health had issued 843 licenses to practice medicine: 411 on the basis of earned medical degrees, 338 on the basis of ten years of continuous prior practice, and 94 by examination. The board did not specify how many aspirants had failed their examinations, though the number was apparently substantial. Reeves also reported that the board's implementation of the law had prompted a large number of physicians to leave the state. While some were fleeing for good, convinced that they would never be granted a license by the new West Virginia board, several others were said to be enrolling in neighboring state medical schools to obtain formal MD degrees so they could return to their practices and be licensed on that basis. Secretary Reeves applauded both of those trends as excellent results of the law; non-Regulars and the poorly prepared were disappearing, while those intending to return with degrees would raise the overall educational level and scientific training of West Virginia doctors.

As Reeves noted in his first annual report to the governor, the vast majority of West Virginia physicians had voluntarily recognized the board's authority and complied with its regulations. The grandfather clause had successfully forestalled widespread protests among a large group of doctors who were happy to receive state-sanctioned licenses without scrutiny of any sort. And according to Reeves, the new law had "greatly improved the *esprit de corps* of the profession" as a whole. He also noted in passing that "among the diplomas offered for verification, were discovered a few of spurious character, which had either been directly bought or obtained upon a nominal examination. These, of course, the Board rejected." But on the whole, Reeves quite rightly considered it truly "remarkable that no serious complaints of violation of the law have reached the Board."[10]

Not everyone, however, was pleased with the board's vigorous exercises of power during the summer and fall of 1881. Implementation of the new license law during those six months had caught the attention of many previously uninterested physicians. Given the nation's long history of refusing to impose restrictive licenses on the practice of medicine, many doctors initially assumed this latest effort would go the way of all those that preceded it. Even after passage of the license law, they doubted that the board would actually challenge them. After all, many of them had practices that provided services to long-time friends and neighbors, often in areas where few alternatives existed. But they now began to realize that a license based on what amounted to AMA criteria—an entirely unprecedented and hence highly unlikely possibility—was improbably and rapidly becoming a functional requirement in their state. If the arch-Regulars serving on the West Virginia Board of Health were allowed to continue implementing their new license law as aggressively as they had been implementing it during their first six months of existence, a substantial minority of physicians began to face the fact that they would be in trouble.

Those newly aroused opponents of rigidly Regular requirements joined forces with the legislators who had opposed any form of medical licensing in the first place as an unnecessary, unjustified, and unjustifiable intrusion into the free marketplace of medical services. Most of those disgruntled delegates were anti-corporate Farmer-Labor Democrats, but a few were Republicans hoping to find issues that might hurt the ruling majority. The *State Journal* of Parkersburg, for example, a Republican newspaper, sarcastically dismissed the Board of Health as the Democratic legislature's way of "instructing people what kind of doctors shall kill them."[11] Those groups, in turn, found additional allies among merchants and local officials who were nervous about the imperious way the board had unilaterally implemented its public health and quarantine powers during the past six months. Together they all saw the adjourned session as an opportunity to curtail—perhaps even stifle—the new state agency before it could be effectively funded and firmly established.

Fortunately for Reeves, he and his colleagues retained the full support of Governor Jackson. Indeed, Jackson opened the adjourned session on January 11, 1882, with an annual message that contained a forceful endorsement of the nascent Board of Health. "The preservation of the public health," he declared, "should be one of the first concerns of government," and he commended the legislators for having done the right thing when they created a state Board of Health during their regular session the previous year. Lawmakers now had a duty to provide that board with adequate funding, he believed, so it could continue the good work already begun.[12]

The board's defenders wasted no time following up. Newspapers reported that friends of the board launched an unusually aggressive and well-funded lobbying campaign on the first day of the session. Reeves in particular pressed lawmakers relentlessly. Other leading physicians of Wheeling pitched in as well, often by arranging generous receptions and private meetings.[13] Under the direction of Baird, "high-minded, progressive physicians representing large and influential constituencies" also renewed their pressure from outside the capital city.[14] Barbee, dubbed "the medical senator" by the press, worked the floor of the state senate, while the cadre of doctor-delegates, minus the gravely ill Isaiah Bee, did the same on the floor of the house.[15] Reeves would later acknowledge their "united and fearless efforts" in preserving the board's legal authority "to elevate the standards of medicine and surgery in West Virginia."[16]

As he had done in 1881, James Ferguson again managed the procedural aspects of the board's legislative affairs. Since funding was the board's principal goal this time around, Ferguson launched his efforts in the house, where money bills had to originate. Because the legislature was officially revising the West Virginia code, Ferguson chose the tactic of proposing a new bill aimed at "amending and re-enacting" the code's recently added chapter "concerning public health," which was the law that had created and empowered the state board a year earlier. That was a potentially dangerous tactic, since defeat of the substitute bill would jeopardize the legal status of the board. But Ferguson knew what he was doing, and the substitute bill not only clarified awkward wording in the original act but also inserted a section authorizing an annual allocation of tax money. The amount to be allocated was left blank in the early drafts of the new bill, suggesting that Ferguson and Reeves were prepared to dicker with their opponents over how much the delegates would be asked to commit.[17]

Opponents of the board were not content, however, to accept the agency's powers as previously enacted and debate only the amounts of money it would receive in the future. Instead, they returned immediately to the issue of medical practice itself. Clearly, some of them had heard from disgruntled practitioners who were waking up to the fact that the new board meant business. One delegate in particular, an attorney named Emanuel W. Wilson, attacked the stipulation that every member of the state board had to "be a graduate of some reputable medical college." On the floor of the house, he praised "the man who worked himself up by dint of his own industry and native intellect," as Wilson himself had done, and expressed resentment of unjustifiable discriminations against "those who had not the advantages of an early education." The license law as previously passed was "undemocratic and unjust" in his view, so he moved to strike the requirement of a medical degree for board membership and also to substitute more lenient terms for automatic licensing.[18]

D. Q. Steere, one of the doctor-delegates, rebutted those arguments in a heated exchange punctuated by interruptions and snide comments. Perhaps the public could tolerate self-taught lawyers like Wilson, Steere asserted, because their actions affected "only your pocket-books." But the public could not afford physicians who learned on the job. "The man who rises to eminence in medicine or surgery without special training for the grave duties of his profession, mounts to that position over the dead bodies of your wives and children." Moreover, the public would not be well served by letting "a man with only knowledge of a few herbs" be eligible for the "responsible duty of examining candidates for admission into the ranks of the medical fraternity."[19]

Ferguson then pointedly expressed frustration with lawmakers like Wilson, who insisted on reraising questions already settled in the previous session, but he patiently defended the revised version of the Board of Health Act. He pointed out that the new language made at least two significant concessions to those who saw the whole operation as a monopolistic attempt on the part of the MSWV to drive non-Regulars from the marketplace. First, while the governor was still required to appoint only "graduates of reputable medical colleges" with twelve years of continuous practice to the Board of Health, the bill's sponsors had dropped the clause that board appointees needed also to "have distinguished themselves by devotion to the study of medicine and the allied sciences." That concession was made because several opponents took the original language to imply that active MSWV membership would be a de facto requirement for board appointment.

Second, where the original law guaranteed a license to anyone with an MD degree from "a reputable medical college," the measure now before the house guaranteed a license to anyone who graduated from "a reputable medical college in the school of medicine to which the person desiring to practice belongs."[20] The expanded language not only acknowledged the existence of different approaches to medical practice but even opened the possibility of automatically licensing graduates from non-Regular colleges. But the crucial word "reputable" was still there, and the all-Regular board would still be determining what constituted a reputable MD degree, regardless of who conferred it. Indeed, the added phrase may have been deliberately disingenuous, as subsequent actions of the board would later suggest. But for present purposes, the specific recognition of different schools of medicine damped the fire of several opponents who had openly resented what looked to them—with good reason—like the first stages of a state-sanctioned takeover by the Regulars.

Two other MSWV allies in the house also made strong speeches in support of the bill. One argued that its opponents had "grossly exaggerated" its likely impact, and the other argued—apparently with a straight face—that passage of the bill would actually work against the self-interest of the medical profession because "whatever tended to preserve or improve the public health must curtail the physician's practice."[21] Unpersuaded, a hostile delegate then demanded a vote to strike the educational requirement altogether, but his motion lost badly. Another skeptical delegate tried to reduce the number of years of practice required for licensing under the grandfather clause, but that also failed. A third delegate proposed automatic licensing for anyone who "attended at least one session of some reputable medical college," whether or not they had been awarded a formal MD degree, but that too lost.[22] Though the board's aggressive implementation of its licensing powers during the second six months of 1881 had aroused more opposition in the adjourned session of January 1882 than the original bill had faced a year earlier, Reeves's allies in the lower house were holding their own in defense of the law's key provisions.

When debate finally shifted from licensing to funding, protracted haggling resulted in an annual salary of $1,000 for the board's secretary and an annual appropriation of $3,000 to underwrite the board's future operations. Delegates then discussed the vexing question of who would pay to execute sanitation orders issued to local jurisdictions by the state Board of Health. Those discussions produced a clause in the revised bill that put financial burdens of that sort on the separate county health boards in the areas affected, not on the state Board of Health or on the larger state budget. With those funding clauses in place, the bill as a whole passed its third reading by a vote of 39 in favor, 14 opposed, and 12 absent and not voting. Ferguson promptly sent that result over to the senate chamber.

Virtually every newspaper in the state expressed approval of the house vote. For every one of them, however, the key issue was public health, not medical licensing. Just as Reeves had intended, the general public saw the new Board of Health first and foremost as a state-level guardian against epidemic disease, not as the supervisor of individual physicians. This perception was almost certainly a result of the quarantine imposed at the board's first meeting back in June. Whether or not the quarantine was fully justified at that time, smallpox had subsequently broken out in epidemic proportions downriver in Cincinnati. Since the disease remained largely absent from West Virginia, the press hailed the new state agency as a "savior" because no systematic quarantine could have been imposed on all the state's local jurisdictions prior to the board's creation. And since that quarantine was still in effect in Wheeling when the legislature assembled in January, everyone in the capital city remained well aware of the board's presence.

Even the Republican *Wheeling Register*, normally no friend of the overwhelmingly Democratic legislature, lauded house passage of the board bill and called for senate concurrence "as fast as possible." The state was "threatened with an invasion of smallpox, and nothing should be left undone which promises safety to the people. . . . But for the energetic execution of the present law, Wheeling, by this time, would have been overrun with smallpox," argued the editor, "her business interests greatly damaged, and the Legislature scared away."[23] The board's bold and preemptive policies on that front were thus paying handsome dividends. The Martinsburg *Independent* was equally delighted that West Virginia finally had state authorities in place to coordinate defensive measures against smallpox, "strike where it may within our borders."[24]

The revised board bill came up for consideration in the state senate in early March. Members debated the measure for the better part of two days, but Reeves was delighted to see his opponents focus their attention on funding issues rather than the licensing provisions. The latter seemed intact after the unsuccessful assaults in the house, so opponents in the senate decided that fiscal strangulation offered the best chance to minimize the power of the new agency.[25] They struck first at Reeves personally, cutting the secretary's proposed salary from $1,000 to $500, and then reduced the board's proposed annual appropriation from $3,000 to $1,500. But defenders of the new board, led by Barbee on the floor and Reeves in the lobby, parried more serious thrusts that would have limited the board's statutory authority. In the end, opponents settled for an understanding that the board's secretary would not have the power to issue binding edicts or mandatory orders to local jurisdictions without concurrence of the full board. That was another direct slap at Reeves, whom some regarded as a man fully capable of becoming a law unto himself under the pretext of protecting the public health. With those clarifications, the bill passed the senate.

When the senate's amendments returned to the house, Ferguson and the doctor-delegates accepted the reduced funding but balked at some of the restrictive language introduced by the senate. Confronted with their strong house response, the senate in turn reworded some of the clauses most objectionable to the board's allies, particularly those aimed at Reeves in such a thinly veiled ad hominem manner. The ultimately reconciled bill was then repassed in both chambers, roughly a year after its initial passage by the same lawmakers. The key final vote in the house of delegates, taken March 15, 1882, was 44 in favor, 7 opposed, and 14 absent and not voting.[26] As a result of that vote, the State Board of Health Act—with the licensing provisions Reeves had originally drafted still intact—became a permanent part of the revised code of West Virginia.

In recognition of his achievements, the MSWV elected Reeves president in 1881, thus putting him into an unprecedented position of power: head of both the state Board of Health and the state medical society simultaneously. In his presidential address to the latter, he reiterated his surprise and delight that the lawmakers elected to the 1881 West Virginia legislature had first passed and then adjusted and repassed exactly the kind of law he had hoped for since organizing the MSWV in 1867. Those legislative achievements had not been easy, he reflected, because "the *genus* demagogue was also ably represented" among the delegates elected to that legislature. They made "the most *pathetic* [his emphasis] appeals," he continued sarcastically, "in defense of the rights and liberties of the dear people against the 'closed corporation'" of the MSWV; they resisted the "'high-handed despotism' of health law"; and they struggled "to save the State Treasury from paying tribute to the offended goddess HYGEIA!"[27] But in the end, he and his allies—both inside and outside the MSWV—had emerged triumphant.

Reeves regretted that political reality had necessitated the ten-year clause. Unfortunately, he conceded, his board had no choice but to license many in "the ten years' class" who were "not qualified to treat the sick," including a number of physicians who considered themselves Regulars. He chalked that off as a political price worth paying, however, because it was a temporary loophole that would never reopen. He then left no doubt about what he intended to do with the unprecedented authority delegated to him and his Board of Health: "Since the 8th day of March 1881," he proudly informed his MSWV colleagues, "there have been no additions to the profession except by the college diploma, or examination by the board."

Reeves was clearly proud of what he and his allies had achieved. "Though one of the younger states," West Virginia had seized "the high ground" with regard to upgrading American medical practice. The legislature had enacted what he characterized as a unique "model of law," a statute containing both licensing provisions and criminal sanctions that AMA Regulars in other states all across the nation could envy in the short run and try to emulate in the long run. Moreover, West

Virginia's new Board of Health had begun to actually exercise its powers and enforce the law. Recognizing and acknowledging those achievements, AMA delegates elected Reeves to a term on the AMA's national Judicial Council at their annual convention in 1882. Reeves, in turn, reconfirmed his unwavering belief that the future of American medicine depended upon the creation of more licensing boards like his in West Virginia, boards with the legal authority to maintain the high standards promulgated by the national organization.[28]

CHAPTER SEVEN

The Dents Confront the Board

A few days after passage of the 1882 revised Board of Health Act the *Wheeling Intelligencer* devoted its lead editorial to the issue of medical regulation and medical practice. "As is well known," it began, "a very determined effort has been made in this State, and with a fair degree of success thus far, looking to the raising of the professional standard for the practice of medicine. Up to the consummation of this movement," it continued, "West Virginia was one of the States in which every quack could settle or sojourn and impose on the public. As matters stand now, a practitioner must show proper credentials from the medical examiners before he can, except at his peril, attempt to practice medicine." But the *Intelligencer* still claimed to be worried because, as the editor warned, plenty of would-be physicians were likely to present educational credentials that were not really "proper."

The editor, who had strongly supported James Reeves and his allies through both the regular session and the adjourned session of the 1881 legislature, went on to characterize many of the nation's medical schools as little more than frauds. Too many of them functioned, in his words, as nothing more than profit-seeking "college corporations" established for the purpose of exacting high tuition payments in exchange for guaranteed but worthless degrees. The "crude and unfitted young men" holding MDs from those schools were flooding an already overcrowded field with additional incompetent practitioners. Consequently, urged the editor, the new board had a duty to watch out for shoddy educational credentials and to reject phony or superficial degrees presented for automatic qualification under the new law. And the public, claimed the editor, would fully and enthusiastically support the board's refusal to accept all degrees at face value.[1]

Reeves himself had almost certainly planted this piece of journalistic cheerleading. He wanted the appearance of public support because he and his colleagues on the board were already engaged in a pair of nasty fights over precisely the issue raised by the editorial: how to determine what constituted a "reputable" degree for the purpose of certification under the new license law. By no accident, both of those tough cases involved members of the Dent family, whom Reeves had come to regard as dangerous subversives. Arthur Dent, in his controversial paper, and Arthur's uncle William Dent, in his presidential address, had both heretically befouled the sacred temple of the Medical Society of West Virginia by articulating a tolerant, inclusive, and somewhat skeptical vision of how American medicine might best evolve. Reeves, now back in a stronger position than ever before and armed with the authority of state law, was determined to show the Dents which vision of American medicine would prevail in West Virginia under his watch.

The first of the two Dent cases involved a request for certification from Arthur Dent, the Weston physician and radical freethinker who had the temerity to suggest at the 1879 MSWV convention that much of what passed for Regular scientific medicine had little or no demonstrable value. When the first Board of Health law went on the books in 1881, Arthur did not have a formal medical degree. Seven years earlier, he had attended lectures at Starling Medical College in Columbus, Ohio, where his uncle William had earned an MD before the Civil War. Like many other physicians of that era, however, Arthur had not stayed long enough to earn a formal degree before returning to establish his Weston practice. Nor had he been practicing for a full ten years in West Virginia since his return. Consequently, because he was neither a degree-holding MD nor a ten-year practitioner, Arthur presented himself for certification by examination, just as George Garrison, Seymore Hopkins, and several others before him had done.

On January 13, 1882, Reeves and Moffett—the two members of the state board from Arthur's congressional district—administered the examination. It was partly written and partly oral. Given the medical views Dent had expressed two years earlier at the MSWV convention in Martinsburg, insiders were not surprised to learn the next day that the two examiners flatly rejected Arthur's answers as unworthy of a license. In a subsequent letter, Reeves enumerated a number of the errors that Arthur was alleged to have made during the oral part of his examination, most of which involved Latinate terminology associated with various bones, muscles, organs, and procedures, as distinguished from actual practices or appropriate therapies. Put differently, Arthur Dent seems to have failed what was largely a vocabulary test, as distinguished from a test of his doctoring skills.

Rather than appeal the results of his exam, which he could have done under the law, or leave the state, as Seymore Hopkins had done after his rejection by the same two examiners, Arthur decided to return to school and finish his MD. That would allow him to reapply for a license as a degree holder. His decision to do so was not unusual. The board had already reported with pride that implementation of the new license law was inducing many of West Virginia's nondegreed physicians to do the same thing, which the board regarded as a positive development. So ten days after failing Reeves's exam, Arthur took the train over to Columbus, Ohio, to see about enrolling in the Columbus Medical College. That institution had been founded in 1876 by disgruntled professors from Starling Medical College across town, where Arthur had studied before. Three members of the Columbus faculty already knew Arthur from his earlier stint at Starling, and they welcomed him warmly to their new college. Arthur had arrived in Columbus just as the courses he needed for a degree were nearing completion, but his old friends on the faculty allowed him to enroll late and take his chances on the exams. Arthur did just that and passed all six of his exams, according to the various professors involved, with solid scores.[2]

The Columbus faculty then granted Arthur credit for the work he had previously done at Starling, recognized his intervening years of experience as a de facto apprenticeship, and noted that his "regular professional standing" had been formally affirmed by "two reputable physicians of West Virginia." The Columbus faculty also noted explicitly that the Board of Censors of the MSWV had admitted Arthur as a member of that organization, which ironically would have afforded him an automatic license if the legislature had not struck that clause from Reeves's original bill. In view of all those factors, Columbus Medical College waived the usual full-time residency requirements—a waiver that was not unusual in that era in any event—and granted Arthur Dent a formal MD degree. On the evening of February 24, exactly a month after his arrival in Columbus, Arthur marched across the stage of Comstock's Grand Opera House and received his diploma with the fifty-eight other members of that year's graduating class from Columbus Medical College. Back home, the *Weston Democrat* announced that their townsman had graduated "with honor." And "Dr. Dent," that paper continued, "is so well known to our people, not only for his skill as a physician but [for] his merits as a citizen, that we know all will take pride in his advancement. We trust it will be the pleasure of the

Doctor speedily to return to Weston and resume the practice of his profession here."[3]

Not everyone back in Columbus, however, was proud of Arthur Dent's MD degree. James F. Baldwin, a dissident member of the Columbus Medical College faculty, who would later start yet another breakaway medical college in the city, wrote privately to Reeves claiming that the whole business had been a fix. Baldwin's letter accused the president of Columbus Medical College of arranging to grant a degree to an old student in exchange for new tuition. When the president found out about the letter, he promptly fired Baldwin. Baldwin thereupon published his accusations in a pamphlet, which he sent to several prominent medical journals. The whole mess soon hit the national medical press as the "Columbus Medical College Imbroglio," which made the legitimacy of Arthur Dent's degree the subject of finger pointing and tongue wagging all across the country.

As might be expected under such clouded circumstances, Reeves was furious when Arthur boldly reappeared before his board in early March with his new MD in hand, again asking to be licensed, this time on the grounds that he was a degree-holding graduate. Barely seven weeks had passed since Reeves and Moffett had ruled Arthur unfit on the basis of his examination. Additional accusations regarding Arthur's degree were then hurled back and forth across the Ohio River. Reeves publicly called the standards of the Columbus Medical College "a stench in the nostrils of all corporate bodies and decent practitioners in the entire country," while the Columbus dean publicly defended the integrity of his college. Nor was Arthur's cause helped by the fact that Reeves and Moffett were simultaneously battling at this time to defend their own standards in the adjourned session of the legislature.

The presentation of Arthur Dent's MD degree from Columbus Medical College forced Reeves and his colleagues to establish formal guidelines for determining the "reputability of a medical college." They had no intention of certifying Arthur; they despised him personally, he had failed the examination they had given him, and he now seemed to be mocking their standards by demanding a license on the basis of an MD he obtained in less than a month under what they regarded as outrageous circumstances. But without specific criteria for rejecting the validity of Arthur's MD, the board would be vulnerable to legal challenge; and should the board lose such a challenge, the entire licensing project would be undercut or destroyed. So they went to work drafting standards. As Moffett put it in quasi-religious language reminiscent of earlier MSWV presidential addresses, he and Reeves needed to establish formal criteria capable of "holding a check on the incompetent, who would dare defame the Temple."[4]

As finally worked out and officially promulgated, the "Minimum Requirements of the West Virginia State Board of Health for Reputability of a Medical College" stipulated, among other requirements, proper facilities, both lecture halls and separate laboratories, appropriate "mechanical and scientific apparatus," and a "capable faculty." Reputable schools would also require qualifying exams before allowing students to matriculate, insist upon full mastery of eight core subjects related to the medical sciences, make students participate in dissections, and demand "actual (not merely nominal) attendance [parentheses in original]" at no less than 80 percent of the lectures offered over a period of at least two full years in order to receive a degree. In short, the board's minimum requirements followed AMA national guidelines and looked, especially with their emphasis on laboratory science and extended study, much like the standards already in place at top Regular medical schools. With those guidelines officially in place, the board rejected Arthur's degree as "disreputable" and denied his right to practice in West Virginia.[5]

Following his second rebuff, Arthur metaphorically threw up his hands and decided to do what Seymore Hopkins had done: quit fighting Reeves and leave the state. Arthur moved eighty miles west of Wheeling to Coshocton, Ohio, a coal mining town similar to the towns he knew in north-central West Virginia. From a personal point of view, Arthur's move proved to be a good decision. Ohio, of course, had no licensing system in place, and even if it should implement one in the future, his Columbus degree would surely be recognized. Once settled in Coshocton, Arthur quickly built a prosperous new practice, became one of the leading physicians of the region, joined civic organizations, and was widely respected in the community when he died in 1900. Back in West Virginia, however, Arthur's name was expunged from the membership rolls of the MSWV immediately following his departure, never to appear again. Reeves and his allies had clearly won that round against the Dents.

Behind Arthur Dent's unsuccessful and ultimately abandoned attempts to gain a license from the state Board of Health, however, loomed a second and potentially more serious confrontation with the Dent family. This one involved attempts to gain a license for Arthur's cousin, twenty-seven-year-old Frank M. Dent, the first-born son of apostate MSWV past president William M. Dent. Frank's case arose from a series of tangled and sometimes disingenuous circumstances, and unlike Arthur's case, this dispute would not be fully resolved until 1889, when the United States Supreme Court finally settled the issue.

While still a teenager, Frank Dent had begun assisting in his father's Newburg-based medical practice, preparing himself through apprenticeship to become the fourth generation of Doctor Dents in Preston County. In 1874, at the age of twenty, he had been made a full partner in his father's practice on the basis of what he had learned from five years of helping his father and reading medical texts. For the next seven years, Frank treated patients both jointly with his father and separately, on his own. Together, then, father and son were continuing the prosperous practice originally established in that part of the state by Frank's great-grandfather. To supplement their medical practice, the Dents also ran a profitable drug and dry goods business. Frank married Ida Latimer Frazier on New Year's Day, 1880, and the couple lived in a house of their own next door to Frank's father and mother. The father-and-son physicians were doing well enough to support a handyman from Ireland

The Dent homes and medical office in Newburg. Photograph courtesy of Richard I. Dent.

and a female servant from Virginia, both of whom lived with the younger Dents.[6]

After the 1881 license law passed, the two physicians on the Board of Health from Preston County's congressional district, C. T. Richardson and George Carpenter, duly began certifying doctors in their jurisdiction. On August 4, 1881, they held a meeting in Grafton to enroll qualified physicians from the surrounding counties. One of the first physicians they certified was William M. Dent, Frank's father, who submitted his name under the ten-year clause. They knew, of course, that William had recently been president of the MSWV, and no matter how much Richardson and Carpenter might have disliked the tolerant policies that Dent tried to persuade the MSWV to embrace, they could hardly turn him away as unqualified for a license under the new law. In addition to practicing for more than ten years, William could also have claimed certification on the basis of the formal MD degree he held from Starling Medical College, a well-recognized Regular school.

At that same initial enrollment session in Grafton, Richardson and Carpenter also listed another physician, whom they identified only as "M. Dent, Newburg," as qualified for a license on the basis of having practiced for ten years. They then sent their initial list of approved physicians over to Wheeling for official registration. Once the names were registered, printed licenses would be mailed from board headquarters back to the physicians whom Richardson and Carpenter had approved in Preston County. Since the clerk of the board in Wheeling had neither a full name nor a full address for "M. Dent, Newburg," he sent that license to William M. Dent in Newburg—along with William's own license—trusting William to get the M. Dent license into the hands of its rightful recipient. But arrival of the M. Dent license generated confusion all around, or at least feigned confusion, since William claimed not to know which other member of the Dent family the two board members had intended to register.

To clarify the situation, William wrote directly to Reeves, even though the two were on—to put it mildly—frosty terms. William explained that his son Frank was out of the state seeking relief from debilitating lung problems, so he, the father, was trying to straighten matters out. William wondered whether a clerical error may have omitted a letter on the mystery certificate, and hence that the license was meant for his son and partner, F. M. Dent, not simply M. Dent. Reeves, in turn, wrote other physicians in Preston County to see what they knew about the situation. From them he learned that most of the area doctors assumed that the license for M. Dent had been intended not for William's son Frank, but as a pro forma courtesy for William's aged father, Marmaduke Dent, even though the latter was more than eighty years old, bedridden with paralysis, and long since retired from active practice. Richardson and Carpenter subsequently confirmed that this had indeed been their intention. Reeves, who was disinclined in any event to be generous toward the only president of the MSWV who had ever dared challenge his rigid principles, became convinced that William Dent was trying to take advantage of personal sympathies, professional relationships, and bureaucratic confusions in order to hoodwink the board into licensing his otherwise unqualified son. From that point on, James Reeves and William Dent were no longer on frosty terms; they became open and implacable adversaries.[7]

In the meantime, Frank himself, apparently unaware of his father's correspondence with Reeves, had contacted Richardson and Carpenter on his own. He wondered whether his combination of five years of practice as an apprentice under his father and seven years of practice as a full partner on his own would qualify him for certification under the ten-year rule. Before answering, those two conferred formally with Reeves and the other three members of the full board and then informed Frank officially that he did not qualify on those grounds; the full board was willing to count only his seven years of practice as an independent partner. After hearing that he would not qualify under the ten-year rule, Frank decided to leave the state and practice elsewhere, as his cousin Arthur had ultimately done. Perhaps a different climate might also improve his nagging lung condition, which was almost certainly tuberculosis.

To explore his options, Frank traveled down the Ohio River to Missouri, west through the Missouri River Valley, and overland into Kansas, looking for likely places to relocate. In community after community, however, he discovered what was still the norm everywhere in the country: physicians of all sorts jostling for patients in crowded medical markets. He eventually chose the rapidly growing city of Topeka as a place where he might be able to establish a practice among newcomers. Following a brief stay, however, he quickly realized that his health would be no better on the Kansas prairies than it had been back in West Virginia; in fact, it appeared to be getting worse. Besides, West Virginia was where generations of his extended family had practiced medicine,

Dr. Frank M. Dent, the young physician at
the center of the case. Photograph courtesy of
Richard I. Dent.

where he already enjoyed the confidence of many loyal patients, and
where he ultimately belonged.

On his way home, Frank stopped in Cincinnati. There he decided to
enroll in the American Medical Eclectic College (AMEC) with the intention of earning an MD degree from that school. He knew that the
1881 West Virginia Board of Health Act had included "graduation from
a reputable medical college" as one of the automatic grounds for gaining
a license, so he reasoned that spending some time and tuition at the
AMEC would allow him to return to West Virginia fully qualified. Following a short series of courses and exams, the faculty awarded Frank
Dent the formal MD he was after, and he resumed his trip back home to
Newburg.

The exact status of the AMEC was difficult to ascertain at the time
and remains difficult to ascertain from the historical record. On the one
hand, Cincinnati was known to be a center of Eclectic medical educa-

tion. Colleges there had been graduating Eclectic physicians since the 1840s, and the institution that eventually consolidated as the Eclectic Medical College of Cincinnati endured through the first four decades of the twentieth century. The education provided at the city's Eclectic colleges was favorably regarded by large portions of the public, and Eclectic graduates intermingled with Regular graduates as highly successful practitioners in many parts of the country—the principal difference between them being the rejection of mineral purges and bloodletting by the former, and a consequent emphasis on botanical applications that were frequently belittled by the latter.

On the other hand, Cincinnati's feisty and contentious Eclectic faculties had a history of frequently reshuffling themselves into new corporate entities under new names, including the Eclectic Medical College, the Eclectic Medical Institute, the Physio-Eclectic Medical College, the American Eclectic Medical College, and Dent's American Medical Eclectic College, which had come into being in 1879. Behind most of those reorganizations lay financial disputes and personal vendettas, as distinguished from doctrinal disputes or educational reforms. But the result was confusion about what faculty and what set of requirements were in effect at any given time in any given iteration of these musical chairs. Moreover, at least some of the reshuffling was done in order to create institutional entities on paper that could issue MD degrees to tuition-paying students with minimal rigor and little actual course work. Mainstream Eclectics were embarrassed by the least scrupulous of these institutions, and Reeves later claimed publicly that the AMEC, which awarded Frank's MD, had been flagrantly selling those degrees "at $30 apiece to all applicants." The Illinois Board of Health, though it was operating on different standards, later condemned the AMEC as well.[8]

When Frank arrived back in Newburg, his wife told him that he had received in the mail what she took to be an official license of some sort. Evidently without bothering to check—or at least without serious scrutiny of the alleged license—Frank happily resumed his old practice as if nothing had happened. The family's rivals in the region, however, took a dim view of Frank's actions and doubted that he was licensed at all. Among those local rivals was Thomas F. Lanham, a staunch Regular and AMA member with corporate and banking ties similar to those of many other elite members of the MSWV. Lanham wrote formally to the Board of Health in Wheeling demanding to know the "class" under which Frank had been licensed; Lanham knew that Frank had been denied a ten-year certificate and knew also that he had not taken a board examination. Reeves replied that he had no idea what was going on, but he would try to find out. After further inquiries, Reeves informed Lanham that the board had no record of issuing a license to Frank Dent under any classification. Lanham, in turn, put word around the county that Frank was an imposter, which forced Frank to admit that his wife had conveniently mistaken a certificate from the American Legion of Honor for a certificate from the Board of Health. Lanham, who would later serve on the state Board of Health himself, promptly demanded that Frank either cease practicing as a physician or face criminal charges.[9]

Rather than abandon his thriving practice, however, Frank turned to his contingency plan. In the spring of 1882, he presented the MD degree he had obtained from the AMEC to Richardson and Carpenter and demanded a license on that basis. He felt especially confident, since the 1882 adjourned session of the legislature had very recently expanded the criterion of "graduation from a reputable medical college," by adding the phrase "in the school of medicine to which the person desiring to practice belongs."[10] Particularly in light of his family's increasingly ugly confrontation with the state's most dogmatic Regulars, Frank was perfectly happy to apply for a license this time as an Eclectic physician with a formal degree from an Eclectic medical college, even though he, his father, and his extended family of physicians had always considered themselves Regulars in the past. Richardson and Carpenter dared not rule on Frank's request by themselves, so they placed his application on hold pending consultation with the full board.

Over in Wheeling, the full board was embroiled in the final stages of Arthur's case, and its members had yet to issue their official statement on standards of reputability. Once their formal criteria were promulgated, however, and then invoked to reject Arthur's application, the full board turned its attention to Frank's most recent request. After evaluating Frank's MD from the AMEC against their newly drafted standards of reputability, the full board unanimously rejected his AMEC degree as "not reputable." Furthermore, declared Reeves, he personally intended to invoke the Board of Health Act to compel Preston County officials to indict Frank as a criminal if he continued to practice medicine.

Thus, Reeves, from his commanding positions as secretary of the West Virginia State Board of Health and president of the MSWV, had thrown down the gauntlet. His actions were calculated and deliberate. He realized that his state board now possessed powers unique in the nation, and he knew that his associates in the national AMA, as well as science-and-education Regulars throughout the country, were watching with hope and admiration to see what he and his colleagues might be able to accomplish with their unprecedented authority. Reeves appreciated that attention, and he was determined to show the rest of the nation what could be done. As he boasted in a speech delivered at the annual national convention of the American Public Health Association in Indianapolis a few months after rejecting Frank's application, "only by [the exercise] of such very great authority as is given the West Virginia State Board can . . . reform be made effective."[11]

All sides recognized that the board's ruling against Frank was not only an explicit challenge to Frank's own career but also an implicit rejection of the inclusive and tolerant medical philosophies that he and the other members of his family had championed. When in doubt, Reeves and his colleagues were going to interpret the word "reputable" in ways that conformed as closely as possible to AMA standards, and they were prepared to drive as many nondegreed practitioners and as many non-Regular physicians as they could from the medical marketplace in West Virginia. What made that situation especially perilous for physicians like the Dents was the fact that Reeves's actions were now cloaked with the legal authority and criminal sanctions of the state.

Given both the high personal stakes and the significant professional implications involved in Frank's rejection, the Dents concluded that the time had come to confront the "very great authority" that Reeves had persuaded the legislature to delegate to the Board of Health. After a family consultation, they decided that Frank would continue to practice medicine, thereby daring Reeves to respond. If Reeves made good on his threats by pressing for Frank's arrest, Frank would plead not guilty, thereby setting up a test case with which to challenge the Board of Health law. Even though Reeves and his allies had gained their legal authority from the legislature, the Dents believed they could undermine the board's power through the courts.

For their strategy to succeed, the family needed an effective lawyer whom they could trust. They turned immediately to Frank's cousin Marmaduke Dent. Five years older than Frank, Marmaduke Dent was Arthur Dent's younger brother, the second son of Frank's uncle Marshall. Frank and Marmaduke had been close since boyhood. When

Reeves first heard about the Dents' legal challenge, he had blithely characterized Frank's choice of counsel—in an official report—as a mere "kinsman" of the Dents, "an obscure lawyer . . . with large pretensions," a man of "bad spirit and coarse language," whose "malevolent" arguments exuded little more than "ignorance of the principles of law."[12] Though Reeves thus tried to belittle Marmaduke as an ill-informed country bumpkin trying to play the part of a big-time professional, Frank's cousin was in fact one of the sharpest young lawyers in the state—far more than just a convenient relative willing to take the case.

In 1870, Marmaduke Dent had earned the first baccalaureate degree granted by the newly created University of West Virginia, and in 1873, he had also received the first master's degree conferred by that university.[13] Known for his classical oratorical style, Marmaduke was already emerging as a champion of the poor and a critic of the powerful. Though only thirty-three at this time, he was rising quickly through the ranks of the anti-monopoly wing of the Democratic Party. Known throughout his career for brave battles in behalf of women's rights, corporate regulation, and racial tolerance, Marmaduke would later embrace the Populist Party. In 1892, ten years after taking his cousin's case, Marmaduke would be elected—as the joint nominee of the Populists and the Democrats—to a seat on the West Virginia Supreme Court. During his twelve years on that bench, he would become arguably the most famous and influential jurist in West Virginia history, and the subject of a 1968 biography by the prominent legal historian John Philip Reid.[14] In short, Reeves's sneers notwithstanding, Frank would be well represented in the proceedings that followed over the next seven years.

The first step in the Dents' challenge went completely as planned. Frank continued to see patients through the spring and summer of 1882. On August 18, at Reeves's behest, the state's attorney for Preston County, a Republican appointee named Neil J. Fortney, filed a series of indictments against Frank for practicing medicine without a license from the state Board of Health. Frank responded publicly in a letter to the *Preston County Journal*, stating that he had "graduated after having attended two full courses and have my diploma from a College whose curriculum is not excelled, and have complied with the law on my part." And the friendly local editor reminded his readers, "An indictment is not conviction by any means. An indicted person may be entirely innocent, as it often turns out on the trial."[15] Marmaduke promptly moved to quash Fortney's indictments on the grounds that the law in-

Marmaduke Dent, Frank Dent's lawyer. West
Virginia and Regional History Collection, WVU
Libraries.

voked by the prosecutor was inconsistent with the Fourteenth Amendment to the United States Constitution—an argument he would eventually carry all the way to the United States Supreme Court. The Preston
County grand jury, however, found otherwise and ordered the action to
go forward.

Subsequent discussions reduced the indictments to a single charge:
the misdemeanor of treating one specific patient on one specific day.
This was probably done for the sake of legal clarity, since both sides realized that this action was intended from the outset to be a test case.
Both sides then agreed formally to the basic facts: Frank conceded that
he did indeed practice medicine on that patient that day without a certificate from the board, and the state's attorney acknowledged formally
that Frank held "a diploma from the American Medical Eclectic College
of Cincinnati." With those understandings confirmed, the case was

scheduled for trial at the April 1883 session of the Circuit Court for Preston County.[16]

The actual trial before the circuit court in Kingwood proceeded in a perfunctory and almost rehearsed manner. Following arraignment, Frank again conceded the facts and pleaded not guilty on the grounds that he had been arrested under an unconstitutional law. Since the grand jury had already ruled otherwise, the trial jury took almost no time in finding Frank guilty as charged. He had violated Section 15 of the Board of Health Act, the punitive section that created the new crime of practicing medicine without a license. The process was over so quickly that the judge granted Marmaduke's request for a short delay in sentencing, so the latter could finish preparing his bill of exceptions. The exceptions were needed as the basis for an appeal.

The next morning, April 12, 1883, the local judge levied a $50 fine against Frank, the lowest amount permitted under the law, and assigned him no jail time. The trial had thus produced what might be labeled a clear and technically unambiguous verdict, on the one hand, without rendering Frank potentially liable for dire pecuniary or personal punishments, on the other hand. That, as the local newspaper noted, was exactly what Marmaduke wanted.[17] He moved immediately to appeal his cousin's conviction to the West Virginia Supreme Court of Appeals, the state's highest tribunal. There Marmaduke knew he would be permitted to confront more fully and extensively what he regarded as the unconstitutional powers delegated by the legislature to the new state Board of Health, powers that Reeves was now exercising in what the Dents and others like them considered to be an arbitrary, heavy-handed, and vindictive manner.

PART FOUR

The Courts

CHAPTER EIGHT

The West Virginia State Supreme Court

I mmediately following Frank Dent's conviction in the Preston County Circuit Court on April 12, 1883, Frank's attorney, his cousin Marmaduke Dent, filed a writ of error that would allow him to appeal the Circuit Court verdict to the Supreme Court of Appeals of West Virginia, the state's highest tribunal. His appeal was based on the grounds that the "act of the legislature passed March 15th, 1882, styled an act 'concerning public health,' was unconstitutional and therefore void, so far as it interfered with the vested rights of this defendant in relation to the practice of medicine."[1] Marmaduke had intended all along to stage his major confrontation with the Board of Health before the state supreme court, so he was delighted when the justices agreed to put Frank's case on the docket of a special session they planned to hold during June 1884.

While the Dents awaited their date with the state's top judges, they prepared a petition to the West Virginia legislature, which they circulated throughout the state in an effort to build popular support for their cause.[2] The petition urged repeal of the Board of Health law on five grounds. First, such a law was unnecessary. West Virginians were "sufficiently intelligent to choose their own means of employing remedies without restrictions and restraints." Second, the law was a sham. Its real purpose had little to do with public sanitation and everything to do with furnishing "the 'charlatan' who lobbied it through the Legislature means of subsistence as one of its officers." Reeves must have seethed with fury when he read that line, knowing that it was being circulated to all sorts of people all around West Virginia, and possibly beyond. Third, the law was ex post facto, because its punitive Section 15 had the effect of punishing people for actions that were not wrong when the law passed.

The fourth reason would eventually become the chief argument in Marmaduke's formal appeal: the Board of Health law violated the Fourteenth Amendment to the United States Constitution. Drafted and ratified during the tumultuous Reconstruction period of postwar policy making, the Fourteenth Amendment had been principally designed to afford full citizenship and civil rights to former slaves. Following the Civil War, ex-Confederates had realized they could no longer hold African Americans as chattel, since the Thirteenth Amendment ended slavery in 1865. But to blunt the impact of emancipation, white legislators who remained in place at the state level in the South after the collapse of the Confederacy had begun passing laws known as Black Codes. Those laws would have created a legal underclass—a sort of American apartheid—by denying freed blacks a number of rights and privileges afforded to white citizens, including the right to assemble in public, the right to hold some types of property, and, significantly, the right to practice certain occupations.

Outraged by the Black Codes, Northern policy makers at Washington countered first with the Civil Rights Act of 1866 and then with the Fourteenth Amendment, which was ratified in 1868. That amendment's famous first section, arguably the single most important section of the entire Constitution, conferred full citizenship on "all persons born or naturalized in the United States" and forbade state legislatures to deny any citizens their full constitutional rights: "No State shall make or enforce any law which shall abridge the privileges or immunities of citizens of the United States; nor shall any State deprive any person of life, liberty, or property, without due process of law; nor deny to any person within its jurisdiction the equal protection of the laws." Arbitrarily singling out specific individuals or specific groups for unequal treatment would henceforth be unconstitutional, and the federal government would have the right to intervene against any such unequal treatment meted out by policy makers or governmental officials in the separate states.

The Dents were convinced that the state of West Virginia had violated that first section of the Fourteenth Amendment when its lawmakers authorized the state Board of Health to bar citizens like Frank from continuing what had previously been a perfectly legal practice. The state did not—and could not—demonstrate that persons holding an Eclectic medical degree, for example, were less effective healers than persons holding a Regular medical degree. Indeed, the ten-year certification tacitly conceded that veteran Eclectics and many others even less well trained could safely be tolerated. Thus, they argued, the 1882 law was not a fair-minded measure designed to protect the general public; it was a vehicle whereby one group of physicians with one approach to the current and future practice of medicine had commandeered the authority of the state to arbitrarily exclude from the profession any physicians with different approaches to the current or future practice of medicine. Consequently, the arbitrary powers being exercised by the Board of Health violated both the due process provisions and the equal protection provisions of the nation's highest law. To support this argument, Marmaduke quoted sections of a recent decision by Judge Anson Willis of the District Court for Washington City, which opposed the imposition of professional standards on those same grounds.

The fifth and final point in the Dents' public petition reminded the legislature that the general public had never asked for medical regulations of any kind. And far from welcoming the new medical license requirements, the petition asserted, most ordinary people in West Virginia "looked on them as an infringement upon their natural and 'inalienable rights' in the interest of a favored class." Laws designed to benefit special interest groups—especially when they lacked both popular demand and public support—were not only misguided but unconstitutional. This one should be repealed.

No record has been found to indicate how many West Virginians signed this petition, but it attracted enough attention to elicit a blistering personal reply from James Reeves. Reeves characterized the Dent petition as the work of a few would-be physicians whose "false pretenses" had been "expos[ed] to professional ridicule and contempt." The increasingly desperate efforts of those would-be physicians "to denounce the 'tyranny' of the State Board of Health, and to speak unkindly of the Secretary" had reached "a climax of hate" in their "vulgar" and dissembling effort "to mislead public opinion."[3] Public accusations and counteraccusations aside, all parties realized that their most significant showdown to date was now scheduled to take place in the chambers of the West Virginia Supreme Court of Appeals on June 17, 1884.

The Supreme Court of Appeals, as mandated under the West Virginia constitution of 1872, met annually in Wheeling for two terms and then held any special terms it considered necessary. In 1884 the court comprised four justices.[4] All four were openly avowed Democrats, like the Dents. Thomas Green, the senior justice, had served since 1875; Okey Johnson, since 1877; Adam Snyder, since 1882; and Samuel Woods, since 1883. The first two had been elected in partisan contests to staggered twelve-year terms in accordance with the state constitution; the last two had been appointed by Governor Jacob Jackson—close ally of James Reeves and a strong proponent of the Board of Health—to fill the unexpired terms of judges who had recently died.[5]

When formal arguments began, Marmaduke worked from an eighteen-page brief that was subsequently published separately for public circulation. Echoing some of the rhetoric of his legislative petition, he began by addressing general philosophies of republican democracy. Harking back to early American constitutional theory, he made the case for limited government and popular authority. "When the Legislature or any branch of government exercises any power not expressly given or implied of necessity, it is guilty of usurpation." Consequently, he argued, since no one could demonstrate a clear and compelling need "to regulate the practice of medicine" in the manner prescribed in the Board of Health law, the act of doing so was prima facie unconstitutional in the broadest sense and should be summarily rejected by the court.[6]

By ratifying their state constitution, Marmaduke continued, West Virginians had "formed a compact for mutual protection of and against each other," and "no man nor set of men under the pretence [sic] of making, declaring or executing law should be allowed to vary or violate that compact" in the manner that Reeves and his allies had persuaded the legislature to let them do. Lawmakers who went along with such schemes were either dupes, incapable of recognizing their democratic responsibilities, or de facto coconspirators, "guilty of betraying those who trusted them." Foreshadowing arguments that he would later make as a Populist judge sitting on the same bench he was now addressing, Marmaduke warned that government was slipping into the hands of self-serving obfuscators. "There is too much law-making," he asserted. "Complexity and consequent corruption is fast taking the place of early simplicity. Every patriot should be on his guard" against those who would deploy the powers of government for their own purposes rather than the common good.

The brief continued with a short disquisition on the meaning of liberty, which Marmaduke presented—in the spirit of the founding fathers—as an expansive concept rooted in protection of individual rights against unmerited impositions from the state. The Board of Health law was manifestly inconsistent with those broad definitions of liberty and in conflict with both the West Virginia constitution of 1872 and the state's own bill of rights. He pointed out that those documents had been the basis for restoring the right to practice law to West Virginia's former

Confederates, a liberty that Republican administrations had tried to deny them in the immediate postwar period. If lawyers enjoyed a constitutional liberty to continue the practice of law, a recognized profession with the right to determine its own membership, then surely, Marmaduke reasoned, Frank Dent enjoyed a constitutional liberty to continue the practice of medicine, which was not a legally recognized profession and did not have the right to determine its own membership. "Every man has the natural right to be a physician and practice medicine subject only to such reasonable restrictions as may prove necessary for the general welfare of the public."

Marmaduke realized, of course, that defining "such reasonable restrictions as may prove necessary" lay at the heart of his dispute with the board. So he did not attempt to argue that state lawmakers had no right to impose restrictions for the protection of the public; clearly, under their so-called police powers they did. Earlier state legislatures had exercised those powers in justifiably reasonable ways, such as making it a crime to perform an operation while drunk.[7] But he argued that the state legislature had no right to impose restrictions that could not be justified, and no right to delegate the judgment of what might be justifiable to "an arbitrary board." Employing a metaphor long popular in American legal discourse, Marmaduke declared that no one in West Virginia should ever be "at the mercy of a star chamber determination of what might be a corrupt and venal board chosen from political or other personal motives by the State Executive." Such a process circumvented the fundamental American premise of open justice. Indeed, Marmaduke proclaimed, again sounding like a member of the Revolutionary generation, even a monarchy "would be preferable to the decree of these arbitrary boards, the stool pigeons of executive power."

If the legislature could somehow determine "the *exact quantum* [his emphasis] of such acquirements necessary" to provide a given service to the public in a safe manner, the legislature would be justified in requiring them. But in the case of medicine, no such "acquirements" could be determined. "Education," Marmaduke acknowledged, "is a great help, but it is only a means of development, and no education can supply a natural deficiency nor create a faculty where none exists." The vast number of extremely effective physicians without formal degrees, coupled with the vast number of formally educated physicians who "make the most absurd diagnosis," offered prima facie proof of that proposition.

To illustrate the fallibility of supposed experts, Marmaduke cited the egregiously embarrassing performance of President Garfield's elite team of doctors three years earlier. The president had been shot by an assassin but did not die immediately from that wound. Everyone in the courtroom remembered the agonizing eighty days that followed, as the president's physicians—notwithstanding their impeccable credentials—repeatedly probed for the missing bullet with dirty fingers and dirty instruments, inadvertently punctured the president's liver, and almost certainly introduced the infection that finally took his life. Congress had been so disgusted with the president's medical care that they refused to pay the bill later tendered by the physician in charge.[8]

Marmaduke turned next to the three provisions under which a person could be licensed to practice. Requiring possession of a formal medical school degree might conceivably be "endurable," he surmised, had the legislature also set aside a fund that enabled anyone who aspired to a career in medicine the financial means to enroll in medical school. But the legislators had never even considered such a thing. Therefore, the criterion of a formal MD degree was both unjustified on its face, since degrees did not ensure best practice, and discriminatory in the real world, since the doors to a medical career would be effectively closed to the able poor. The latter would be a great loss to humanity at large and to West Virginia in particular, especially since the state had no medical colleges.

Marmaduke then addressed the ten-year rule, arguing that its inclusion in the law paradoxically supported his client's case, not the state's case. That provision demonstrated that the old system was not so harmful that it needed to be scrapped for the safety of the public. Instead, the ten-year provision merely confirmed the political nature of the whole project. The cabal behind passage of the law realized they could not risk alienating the vast majority of physicians already practicing in West Virginia, and they dared not face an angry public suddenly deprived of hundreds of doctors whom the pubic continued to trust. So Reeves and his allies had introduced the blatantly awkward grandfather clause, thereby undermining their own assertions about the dangers and inadequacies of the medical marketplace already in effect.

In discussing the examination clause, which the state would surely present as a mechanism of fairness and opportunity, Marmaduke returned to the theme of arbitrary power. "This provision is meaningless, except that it places the liberties of a large number of our fellow citizens under the control of a few grandees whom it clothes with despotic powers." Since the law stipulated no standards whatsoever for assessing the results of any given examination, physicians on the Board of Health were free "to decide what are the necessary qualifications in each individual case, and admit or reject an applicant who does not know his stomach from his brains, or through ill-will, party influence or a domineering spirit [here Marmaduke surely had Reeves in mind], reject an applicant who may know more and be a better practitioner than the combined board."

Most physicians in the state, he pointed out, had begun learning their craft by studying under an experienced practitioner, and only "after a knowledge of the simpler diseases and their remedies is acquired, to engage in practice, probably as a partner with his tutor," just as his client Frank Dent had done. Perhaps later, after gaining practical knowledge and the necessary funds, they might attend a "college course." That tried-and-true route would now be blocked because no one could begin a practice in the first place, even if restricted to straightforward cases and even if under the supervision of a senior physician, unless they were already "*fully qualified in the judgment of the board* [his emphasis]." Marmaduke doubted that lawmakers intended such results, but by delegating so much arbitrary power to the new board, the legislature had created a legal "creature . . . greater than the creator," and that "creature" was now manipulating the law to its own ends.

Toward the end of his general arguments, Marmaduke mocked evocation of the public welfare as "the excuse for the enactment" of such a law. The public had not asked for such protection, and citizens capable of selecting their own legislators and their own public servants were surely capable of selecting their own medical attendants. "The public welfare," he stormed, was "the most abused of all terms and cover for all tyrannies from the beginning of human government. . . . The public welfare burnt the witches in New England, kept the negro in bondage, and shamefully refuses the extension of political liberty to the crowning jewel of God's works, fit to serve us as mother, sister and wife, every way except in the mere matter of civil government."

Most of the people of West Virginia opposed the Board of Health Act, Marmaduke claimed, because it insulted "their capabilities," interfered in "their private matters," and, worst of all, "attempt[ed] to place them at the mercy of a monopoly." Building to a crescendo, Marmaduke pulled no punches. The Board of Health was "purely the work of a few ambitious physicians," a tight-knit subset of the state medical society. Their real purpose had nothing to do with the public welfare. Instead, "the real object is to shut out competition, diminish the ranks of the profession, raise their fees, increase their wealth, and give themselves political prominence." Playing with medical metaphors, he accused Reeves and his allies of "insert[ing] their probe in the body politic" in order to bring forth their own variety of "*laudable pus* [his emphasis]," a reference to the belief among many physicians that the production of pus at the site of an infection was a positive sign that the body was fighting back. They "fastened a few leeches upon the public treasury, although profuse bleeding has lost favor with the profession." They enabled the board's "secretary [Reeves, of course] to advertise himself" as the embodiment of "perfection," while "filling our courts with indictments against practitioners who are resolved not to place their property, their rights and the future" in the hands "of an irresponsible board of illegal despots."

Only toward the end of his largely rhetorical appeal did Marmaduke turn to technical points of law and precedent. He based his formal request to nullify the Board of Health law—and thereby reverse the judgment against Frank—on two basic constitutional grounds. The first involved the law of property and the due process clause of the Fourteenth Amendment to the United States Constitution. Because Dent's medical practice was a form of legal property, the state, acting through an unelected board that answered to no one, did not have the legal right to take it from him without due process. The board's attempt to do so not only flew in the face of a long history of strongly protected property rights and sustained opposition to state seizure but also directly violated the recently amended United States Constitution. If the 1881–1882 legislature had the right to ban all physicians who had not practiced for the arbitrary length of ten years, Marmaduke suggested, the next legislature could increase "the limit to twenty years, the next to thirty, and so on." Ultimately, if laws like the Board of Health Act were allowed to stand, no property rights would be safe from the arbitrary actions of whoever controlled the state.

In defense of professional practice as a form of property, Marmaduke cited first and most prominently the 1867 case of *Cummins v. Missouri*, in which the United States Supreme Court had effectively restored to former Confederate lawyers in that state the right to practice their profession. Immediately after the Civil War, the Missouri legislature had effectively banned former rebels from several types of public office and several types of professional and quasi-professional occupations by requiring a loyalty oath from anyone who served or practiced in those capacities. By a 5 to 4 vote, however, the Supreme Court justices had decided that the Missouri legislature did not have the right to deprive anyone of their preexisting property—including their right to resume their professional and quasi-professional practices—without due process.[9]

In Marmaduke's view, what the Missouri legislature had tried to do to ex-Confederates paralleled what the West Virginia legislature was now trying to do to physicians like Frank. The new board was categorically, and without compelling justification, depriving physicians like Frank of their preexisting professional property without due process. And surely if the justices were willing to side with former Confederates in that case, which they decided before passage of the Fourteenth Amendment, the court would now side even more strongly with Frank, since every American's right to due process had been strengthened. Marmaduke also cited the 1866 United States Supreme Court ruling in *Ex parte Garland*, which also involved a former Confederate lawyer. That ruling was less well suited to Frank's case because it involved a prior presidential pardon to the individual who was seeking the restoration of his rights, but it nevertheless underlined the principle that an established professional practice was a form of property that state legislatures were bound to respect.[10]

Anticipating another of the state's likely responses, Marmaduke then argued that the right to appeal adverse rulings to the Board of Health—the right that opponents had inserted into the law before passage—did not constitute due process. Quite the opposite, in his view, because the pernicious law in question armed board members with "absolute power to decide that your diploma is not the work of a reputable institution and that your examination is not satisfactory" on whatever capricious or self-serving grounds they wished. "A more despotic exercise of power cannot possibly be imagined." Consequently, the law wrongly empowered members of the Board of Health to "destroy a vested right" that his client "enjoyed at the time it was passed," and to deprive his client of valuable property "by a process rather ministerial than judicial, and wholly different from that which is meant by the judgment of his peers . . . and this too without any, much less a just, compensation, even admitting the forfeiture to be for public use."

The second constitutional ground upon which Marmaduke rested his appeal was a contention that the Board of Health law was "in effect a bill of pains and penalties and *ex post facto* in its operation." Bills of pains and penalties were laws that imposed punishments short of death—without ordinary judicial proceedings—upon persons presumed to be prima facie guilty of prior acts that subsequently became felonies. For all intents and purposes such bills were indistinguishable from bills of attainder, though the latter could include the death penalty. Accordingly, American legal practice had generally subsumed bills of pains and penalties under the broad heading of attainder. Both, in turn, were generally regarded as ex post facto, since they imposed punishment for things people had done prior to their passage.

The United States Constitution expressly forbade both bills of attainder and ex post facto laws, largely because the American colonists had so deeply resented the imposition of punishing measures enacted after the fact by lawmakers in London, against which they had no judicial recourse. Americans had historically insisted instead that all punishments and forfeitures had to result from ordinary judicial procedures, and American courts had never permitted punishment for an action that was legal when it was done, even if a later law made the act illegal. In *Ex parte Garland*, the Supreme Court majority had ruled that a ban against ex-Confederates reentering specific professions was an unjustifiably broad bill of attainder and hence unconstitutional. Marmaduke now took the same line with regard to the Board of Health law.

He contended that depriving a legally practicing physician like Frank Dent from continuing to function as a physician was a punishment, and a serious punishment at that, since it affected not only the individual but also his family and his community. "If the Legislature had declared that all physicians who did not hold a diploma from a reputable medical college or had not practiced for ten years should forfeit and pay a fine of $1,000.00, the court could not have hesitated to have declared such a law a bill of *attainder* and in its nature *ex post facto*." Yet the punishment inflicted by the Board of Health law in actual practice was even harsher than a fixed fee forfeiture. The board deprived physicians like Frank of their preexisting livelihoods, forcing them in essence to abandon what they had achieved and start their occupational lives over again, though they had done nothing wrong. And it did so without ordinary judicial proceedings. "The Legislature may have been deluded into believing the act to be for the public welfare," Marmaduke conceded, "but this fact does not diminish its obnoxious character, and it is not only the duty but it should be the pleasure of the court to decide it unconstitutional, null and void."

Concluding with an oratorical flourish, Marmaduke reminded the four judges of their new state's constitutionally prescribed motto: *Montani Semper Liberi* ("Mountaineers Always Free"). "Among the mountains of West Virginia dwelt free-born, liberty-loving . . . people," he thundered, who were committed to preserving their inalienable rights "free from encroachments of domestic usurpation." West Virginians had stood up bravely and independently during the Civil War against the arbitrary usurpation of power by unconstitutional secessionists at Richmond. Consequently, he was confident that the state's supreme court would now strike down the Board of Health law—another form of domestic usurpation—and permit Frank to continue "to alleviate the sufferings of humanity untrammeled by the arbitrary regulations of unauthorized legislation."

After Marmaduke finished, the attorney general of West Virginia, Cornelius Clarkson Watts, rose to defend the validity of the Board of Health Act. Influential chieftains in the Democratic Party tagged Watts as a future star, and they had engineered his nomination and election as attorney general in the fall of 1880. Sixteen years later, he would be his party's candidate for governor. At this time, however, he was nearing the end of his four-year term as attorney general, having already established an admirable record as an outstanding litigator on behalf of the state. Watts had served in the Confederate army during the Civil War and was hence, somewhat ironically, a perfect example of someone whose right to practice his profession had been restored by court orders—the same thing Marmaduke was now seeking for Frank.[11]

Watts's brief was not published and apparently did not survive in the court's records, but his argument can be deduced from the thorough summary and subsequent opinion offered by the justices themselves.[12] The most contentious and famous cases that Watts had won earlier in his tenure as attorney general had been tax cases, in which various railroad companies had challenged the right of the state to exact certain levies from them. Watts won those high-visibility cases by defending an expansive interpretation of state legislative powers, which he now applied to the question of medical regulation.

Under the United States Constitution, Watts argued, the separate states possessed virtually unlimited police powers over all matters not specifically assigned to the national government. Public health had long been recognized as such a matter. Accordingly, he rightly pointed out, state legislatures had been implementing various forms of sanitary- and health-related policies since the formation of the Republic, including laws that regulated various aspects of medical practice. Since the Civil War, moreover, legislation of that sort—especially sanitary regulations vested in boards of health—had become more common than ever all around the nation. The West Virginia legislature, in other words, was not inventing some radical new doctrine when it passed the Board of Health law; it was acting upon well-established constitutional principles widely recognized throughout the country. The right of state legislatures to regulate specific aspects of medical practice—by placing special conditions on surgery, for example, or on the dispensing of dangerous drugs—had already been challenged and upheld in Ohio, Maine, Texas, Missouri, and Minnesota. Only in Nevada had an actual license law been set aside, and in that case, he claimed, the court had quibbled over phrasing, not over fundamental rights. While no other state boards of health had licensing powers as strong as those in West Virginia, the differences were matters of degree, not of principle.

Watts thus rested his case squarely on the constitutional premise that state legislatures had the perfectly legitimate and repeatedly sustained authority to enact any policies whatsoever that they considered appropriate for the defense of public health and safety, unless the law violated overriding constitutional rights. And Watts strongly asserted that the Board of Health law overrode no such larger constitutional rights because West Virginia legislators had wisely and properly provided equal treatment, equal opportunity, and multiple pathways to legitimate medical practice. The fact that scattered individuals outside the legislature might dislike or disagree with a given policy that duly elected legislators considered necessary, or might think that the policy went too far, was completely beside the point.

After hearing both arguments, the state's supreme court justices considered the case for four and a half months before rendering their decision on November 1, 1884. Writing for a unanimous court, senior justice Thomas C. Green came down on the side of the state. His long and thoughtful opinion was not, however, without genuine sympathy for Frank Dent's situation. Green began by expressing agreement with Marmaduke Dent's general defenses of liberty and property, but gently pointed out that equally valid interpretations of liberty and property could be cited in defense of legislative integrity and policy responsibilities under the American system of government.

Quoting Thomas M. Cooley's influential treatises on constitutional limitations, Green went on to express his belief that "the judiciary . . . cannot run a race of opinions" with duly elected representatives of the people. "Any legislative act, which does not encroach upon the powers apportioned to other departments of the government, being *prima facie* valid must be enforced." And especially germane to this case was another of Cooley's dicta, which Green quoted: "If courts are not at liberty to declare statutes void because of their apparent injustice or impolicy, neither can they do so, because they appear to the minds of the judges to violate fundamental principles of republican government." In other words, some of the justices on the state supreme court apparently agreed with the Dents that the Board of Health Act was poor public policy, but they did not believe they should nullify it in the absence of explicitly unconstitutional provisions.[13]

Citing numerous cases in numerous states, Green followed Watts in establishing the widely acknowledged police power of state legislatures to regulate businesses and occupations for the benefit of the public welfare. He spent several pages discussing lawyers specifically, a profession he saw as a good parallel to medicine. Every state required lawyers to be licensed as a protection for ordinary citizens, "who are necessarily incompetent to a considerable extent to judge of the qualifications of a lawyer." In Green's opinion, "it [was] obvious that the doctor equally with the lawyer required a special education . . . and the community is no more competent to judge the qualifications of a doctor than of a lawyer." In fact, the public was probably less able to assess doctors than lawyers and hence vulnerable "to be imposed upon by quack doctors." State legislators, in other words, could reasonably believe that West Virginians deserved "additional protection against the humbuggery of quack doctors." If state legislators were convinced that this law would advance that goal, they had a perfect right, even a duty, to pass it.

Green's opinion then reviewed at some length the cases in other states where similar issues had arisen. He agreed in every instance with decisions that sustained the legislative regulation of medical practice, and he took five pages to argue that the lone nullifying judge in Nevada had reasoned poorly; his colleague on the Nevada bench had been correct when he said that courts had no right to second-guess the "motives" behind statutory phrasing, since the reasons behind a particular

Judge Thomas C. Green, author of the West Virginia Supreme Court decision. George W. Atkinson, *Bench and Bar of West Virginia* (Charleston, WV, 1919), 7.

legislative phrase "may have been as varied as the different minds of its members." Courts had the right to determine only whether the lawmakers possessed the constitutional power to pass such statutes in the first place, not whether the statutes were the best possible laws the judges could imagine. "It is simply the question of power which we are called upon to discuss and determine. Whether this power was reasonably or unreasonably exercised . . . wise or unwise . . . expedient or inexpedient . . . was up to other departments."

In the case now before them, Green and his colleagues had taken that approach, and they saw no inherent or explicit conflict between the Board of Health law and either the West Virginia bill of rights or the United States Constitution. While legislators "should permit the utmost freedom of action by each citizen," as Marmaduke had so eloquently argued, legislators surely had the power "to restrain action" when they determined, for whatever reasons, that the "comfort, health or prosperity of the community" was at stake. "This power has been constantly exercised by state legislatures; and the courts have . . . universally recognized such power and have held, that acts of the state legislature passed in the exercise of this power were constitutional and valid."

Toward the end of his long decision, Green returned to the sympathetic tone with which he had opened. When Marmaduke complained that the powers conferred on the Board of Health were unacceptably arbitrary and being exercised in a tyrannical manner, he might well be right. "These may be sound reasons why the legislature should have guarded as far as possible in the law against such evils, but they are very poor reasons why no such law should be passed." In every law, there existed the danger that "those entrusted with carrying it out" might behave "in an unjust and oppressive manner." Maybe, in retrospect, the West Virginia legislators should have included stronger provisions than they did to prevent that possibility. Still, the legislators had the power to pass the law they enacted, and if the Dents did not like it, they would have to go back to the legislature to get it changed or repealed. "But while it remains a law," Green concluded, "it must be enforced by the courts." Frank lost.

Reeves gloated publicly. Lest the press fail to cover his victory, he wrote privately to the state's leading editors, pointing out among other things that the appellate court judges had been unanimous in their decision. The editors, in turn, published announcements that the decision had unambiguously sustained "the validity of the act creating the State Board of Health, and conferring certain powers and duties upon it."[14] Members of the Board of Health took the appellate court decision as a triumphant ratification of what they had been doing for three years and vowed to continue their efforts to prosecute anyone practicing without their imprimatur. Outside the state, influential medical journals associated with the AMA, like the *Medical News* of New York, hailed Judge Green and his associate justices for upholding their state's police powers and called upon other states to follow West Virginia's now legally established precedent for regulating medical practice.[15]

The Dents, however, were not yet willing to concede defeat. Marmaduke obtained a writ of error from Judge Green and filed an appeal to the United States Supreme Court, which the Supreme Court was obliged to hear at that time in cases where state judges upheld state laws against claims for relief under federal law or under the federal constitution. Since Marmaduke had consciously rested Frank's defense on the first section of the Fourteenth Amendment, his request for a ruling from the high court was granted.[16] On October 13, 1885, nearly a year after Judge Green's decision, *Dent v. West Virginia* took its place in a long line of cases waiting to be reviewed by the United States Supreme Court. Observers estimated the line to be about three years long. In the meantime, the Board of Health continued its ambitious licensing project with renewed energy and confidence.

CHAPTER NINE

Conflict and Enforcement

The legal challenges mounted by Marmaduke and Frank Dent against the licensing powers of the Board of Health did not deter James Reeves and his colleagues from continuing to implement their agenda. If anything, given the pugnacious personalities and bitter enmities involved, the Dents' legal challenges probably made the board members less compromising than they might otherwise have been. Consequently, even as Frank's case was pending before the state supreme court, the board's members continued to implement their rigidly defined licensing program as aggressively as possible. After Judge Green rendered the state supreme court's ruling in favor of the board, Reeves and his victorious colleagues were more confident than ever that they could achieve their ultimate dream of making West Virginia the great national model for medical licensing on AMA terms.

With that in mind, the board moved steadily during the 1880s toward their ultimate goal of requiring a formal MD degree from a "reputable" Regular medical school in order to obtain a license to practice medicine. In doing so, they were quite deliberately deploying the power of the state to create a new professional class. The practice of medicine would no longer be a flexible undertaking, open to a wide variety of healers of different sorts in a broad marketplace of unlimited consumer choice. Nor would it be open to self-trained physicians who had gained their skills and reputations as apprentices in the traditional manner, even if they claimed to be practicing Regular medicine, unless they passed a rigorously scientific examination that the board made ever harder and harder to pass. The board members wanted medicine to become a learned profession of high status, open only to those capable of earning an advanced degree that involved a rigorously scientific curriculum.

Notwithstanding the arguments they made to lawmakers in 1881 and 1882, Reeves and his allies in the Medical Society of West Virginia—by their own avowals—had not been fighting first and foremost for a conventional public health statute, which then happened to contain a licensing provision; they had been fighting first and foremost for a license law, which they had gotten through the legislature by embedding it in a conventional public health statute. Consequently, even though the smallpox quarantine was their first public action, the board members devoted almost all their early efforts to determining who was "legally qualified" to practice under the new rules, not to improving water supplies or abating dangers to the public health. As Reeves reported to his friend Governor Jacob Jackson, the board had "deemed it necessary to begin its work by executing the medical [as distinguished from broad public health] provisions of the law," and rigorous enforcement of the licensing provision was "the first step" in that process. Only then, as Reeves phrased it, would the board be "ready to begin its purely sanitary labor."[1] The board's licensing program, however, would shortly put its members at odds with Governor Jackson's successor, who was as skeptical of Reeves's project as the Dents were.

During its first eighteen months of activity, the board had issued 843 licenses: 411 by recognizing formal degrees, 338 on the basis of ten years prior practice, and 94 by examination. They had also, of course, rejected an unknown number of applicants, including both A. M. Dent and Frank Dent. Three months after Frank's conviction in Preston County, the board began circulating sample forms of a formal complaint that anyone could fill out and submit to any public authority if they suspected someone of practicing medicine without a license, and the board also sent sample indictments to all county prosecutors and justices of the peace, which they could copy and use if they received complaints. In January 1883, the board even considered asking the new legislature to amend the Board of Health Act to require all justices of the peace and county sheriffs to check the qualifications of anyone thought to be practicing medicine in their jurisdictions, rather than wait for complaints or inquiries. But Reeves eventually decided not to risk "the danger of unasked-for and pernicious meddling" on the part of lawmakers who might dislike the board's aggressive behavior.[2]

The board issued 198 additional licenses in their second year, almost half of which went to residents of adjoining states whose practices extended into West Virginia. Only ten of those were awarded by examination, and just four on the basis of ten years' practice. In subsequent years the number of new licenses fell to an annual handful. In 1884, the board announced that it intended to stop issuing ten-year licenses, hoping the "*crop* . . . has been gathered [emphasis in the original]"—though a trickle of late applications would continue to come in from remote areas through the end of the decade.[3] The number of physicians licensed by examination also continued to fall. Circumstantial evidence in board reports suggests that fewer and fewer applicants bothered to attempt the special exams, and no more than one in four of those who attempted them passed. The board's actions thus confirmed their goal of granting medical licenses almost exclusively to physicians who had earned formal MD degrees from "reputable" medical schools, which the board had already defined in Regular AMA terms, notwithstanding references in the legislation acknowledging the existence of multiple "schools" of practice. The *Wheeling Register*, long a supporter of the board, was correct when it predicted that Dent's defeat before the state's highest tribunal would "stimulate the Board to renewed exertions and good work must result."[4]

From the point of view of those hoping to transform the practice of medicine from an open occupation to a legal profession, the board's remarkably strict and successful implementation of the licensing provision was nothing short of miraculous. George Moffett, then the board's titular president, self-congratulated the group in 1883 for "put[ting] our shoulders to the wheel . . . to elevate the standard of medical practice within our borders by a rigid enforcement of the law against ignorance and quackery." In his report to the governor, Moffett noted proudly that the board had received "many very high compliments from distinguished sources for the manner in which it has performed its difficult labors—notably, for its efforts to elevate the standard of medical education in West Virginia," by which he meant, of course, the standards required for practice, since there were still no medical schools in the state. Using phrases right out of earlier MSWV presidential addresses, he asserted that the board was "holding a check on the incompetent, who would dare defame the Temple, and repressing the charlatan." He also noted that "the greatest harmony has marked our association as members of the State Board of Health," something the Dents would surely have agreed with, however ruefully.[5]

That same spirit of shared purpose and mutual support linked the close-knit members of the board to the larger MSWV. All of the board members had come out of that organization, after all, so no one was surprised when the MSWV passed a resolution at its 1883 annual meeting declaring that "the State Board of Health and the Medical Society should be in perfect harmony." At that same meeting, delegates voted to

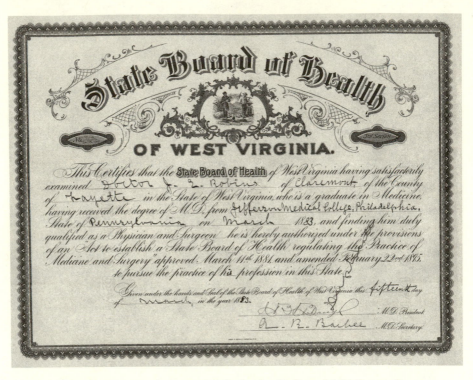

License originally issued in 1883 on the basis of a formal MD degree from Jefferson Medical College. West Virginia State Archive.

pursue the possibility of establishing a joint medical journal with the board. MSWV members further reaffirmed in strong terms their unwavering commitment to strict AMA standards. In yet another resolution, the MSWV conventioneers roundly condemned the Medical Society of the State of New York for its adoption of a policy that opened the door for the admission of Homoeopaths—a policy that prompted the national AMA to expel the New York society—and they called for a boycott against any journal that advocated any sort of consorting with non-Regulars. In his presidential address that year, Benjamin W. Allen, who had recently been appointed to a chair in anatomy, physiology, and hygiene at West Virginia University, hailed scientific education as the only path capable of leading the medical profession to a progressive future, and he lauded West Virginia's Board of Health for following AMA directions without any deviations or degrading detours.[6]

For Reeves personally, the state court's decision in the Dent case also represented the culmination of his own career. Nearly two decades earlier, he had established the MSWV, and for the next fifteen years he had successfully insisted that the new medical society remain resolutely committed to rigidly defined AMA guidelines, even in hard times and even in the face of alternative visions of how best to incorporate science into the practice of medicine. He and his allies had then engineered passage of a law that allowed them to deploy the power of the state against all types of would-be physicians who disagreed with their absolute standards. By 1882, Reeves had risen to official command of both the MSWV and the state Board of Health. His efforts in West Virginia were also gaining him increased national recognition. An 1883 editorial in the *Philadelphia Medical Times* praised "the prompt action of the efficient and active Secretary of the State Board of Health, Dr. Reeves," in driving itinerants and non-Regulars out of his state.[7] The Connecticut State Medical Society made him an honorary member.[8] Members of the nationwide American Public Health Association chose him as their president largely in recognition of what he had accomplished in West Virginia.[9]

Perhaps sensing this culmination, Reeves began telling his closest associates that he had decided to retire from the fray. He had long suffered from asthma attacks, which were recently becoming so severe that he was having trouble maintaining his nearly obsessive pace. Knowing that the mantle of leadership both inside the MSWV and on the state Board of Health would pass to his close ally George Baird, Reeves was confident that both of those institutions would sustain the absolute standards to which he and Baird had always been so passionately committed. That faith freed him to explore the possibility of relocating out of state, in a place where his asthma might be less debilitating.

After touring several possible sites, Reeves eventually decided to retire to Chattanooga, Tennessee. Soon after the state supreme court rendered its verdict against Frank Dent, Reeves began the complicated process of dismantling his practice in Wheeling. In 1885, at the age of fifty-six, he relocated to that Tennessee city, where he would remain until his death from liver cancer in 1896. Though he stayed active in the medical sciences over the next decade and continued to receive national acclaim for his achievements in West Virginia, James Reeves, perhaps the single most influential physician in that state's history, thus voluntarily stepped out of the spotlight and departed the medical jurisdiction he had done so much to reconfigure.

When Reeves announced his retirement to the general public, the remaining five members of the Board of Health passed a formal resolution of gratitude acknowledging him as the man "to whom belongs, in great measure, the credit of directing [the board's] successful labors." In an address to the MSWV, incoming president George Baird offered what sounded like a eulogy to his departing comrade and co-general in the long battle to upgrade the practice of medicine. "Dr. Reeves was the father of the Board," he began, and "it can truthfully be said that it was mainly owing to his persevering efforts that the act creating the Board, and the amended act giving it increased powers were passed by the Legislature." Though forced to endure "abuse and slander," Baird noted, Reeves had sacrificed his own practice to deftly manage the key policies implemented by the board. "The position accorded to West Virginia by medical men in other States as one of, if not the foremost, in creating a State Board of Health, is a tribute to his fostering care." Reeves above all others, in Baird's estimation, had made "the law a success," and his brave, unflinching example would stand as "a model to be copied by other States."[10]

Triumphantly, Baird also noted—in a clear allusion to the *Dent* decision—that the departing board secretary "had the satisfaction of learning but a few days before ill-health compelled him to resign his position that the most malignant of his detractors had been completely overthrown by the highest tribunal in our State." President Baird urged members of the MSWV to continue to honor the work of their founder by never compromising "a strict enforcement" of the AMA's Code of Ethics. Closer attention to those standards "in the past would have protected this Society from the admission of members who proved to be of no credit to it during their membership"—doubtless an allusion to the case of Arthur Dent. As if to ratify Baird's charge, the MSWV Board of Censors then refused to admit a new applicant for membership. While the candidate held a license as the result of passing a board examination four years earlier, the candidate did not have the MSWV's new sine qua non: a formal medical degree. The MSWV's printed transactions pointedly omitted the applicant's name, but everyone concerned knew it was George Garrison.

An ugly confrontation over the choice of Reeves's successor on the Board of Health revealed the resolve of the state's hard-line Regulars, now emboldened by the decision of the state supreme court and by their complete control of the board. In the fall elections of 1884, Emanuel

Willis Wilson had replaced Jacob Jackson as governor. Although both men were officially Democrats, they represented nearly polar opposite wings of that party. Jackson, the corporate attorney who had been born into a family of prominent lawyers, enjoyed the backing of powerful industrial interests. He had actively supported what he regarded as the rational and necessary consolidation of economic and professional power.

Wilson, in sharp contrast, was a self-educated and self-made man who rose to political prominence by attacking monopolies and trusts in an overtly demagogic fashion, first as a member of the state legislature and then as a candidate for governor. While a member of the state legislature, he had actively opposed the consolidation of economic and professional power in any form, and his principal support when running for governor came from discontented farmers and mine workers. Many farmers and mine workers had bolted the Democratic Party in the late 1870s and early 1880s in order to vote for Greenback-Labor alliances rather than support politicians they regarded as puppets of the Kanawha Ring. But they had returned to the Democratic Party in 1884 to vote for Wilson. Of West Virginia's nineteenth-century governors, Wilson was the "most noted foe of corporate privilege."[11]

A proto-Populist himself, Wilson lauded direct experience over bookish learning, and he detested claims of superiority, whether merited or not. Members of the MSWV no doubt remembered that Wilson was the house delegate who unsuccessfully tried to remove the requirement in the then-pending board of health bill that every board member would have to "be a graduate of some reputable medical college." Wilson had also tried unsuccessfully on the floor of the house to substitute more lenient criteria than a formal MD degree as grounds for automatic certification to practice.[12] Formal degrees, in his view, were no more than intellectual pretense; real worth was demonstrated only through practical achievement.

Now as governor, Wilson had the power to fill vacancies on the board whose creation he had opposed as a lawmaker. Reeves's resignation left an opening in the first congressional district, in addition to the regular rotation of two other staggered terms. Rather than nominate replacements, Wilson dragged his feet for more than a year, and some observers feared he might be trying to kill the board by depriving it of members. Their fears increased when the governor also delayed paying the board's expenses. At a meeting of the Council of State Boards of Health in Washington, DC, in August 1885, Reeves warned that there might be trouble ahead in his old state. "Unfortunately for us in West Virginia, in all probability for the next three years we cannot look to the Executive as a friend of the Board. There is not that zeal in his compliance with the law supporting the Board which characterized the acts of his predecessor."[13]

Only after two years of deliberate inactivity, and only under intense pressure, did Wilson finally move to appoint a successor to Reeves on the Board of Health. Not surprisingly, the anti-elite governor spurned MSWV insiders; he regarded them as a bunch of monopolistic wire-pullers who were bluffing their way ahead with fancy degrees rather than effective treatments. The first physician he approached declined to serve, protesting that he was not eligible under the terms of the law, which was true. Consequently, Wilson then turned to perhaps the most overtly insulting candidate he could reasonably defend as a qualified appointee under the terms of the law. He appointed George Garrison, the physician recently rejected for membership in the MSWV.

On the one hand, Garrison was known to have been a protégé of the politically powerful George Baird, the man who assumed leadership of the MSWV after his long-time ally Reeves retired. Baird had once seen to it that his young colleague was among the first three physicians in the state to be licensed by examination, and Garrison had succeeded Baird as the popularly elected health officer of Wheeling. The two had been so close in the early 1880s that Garrison named his son in honor of his powerful patron. On the other hand, Garrison's inflated ego and un-principled ambition had since alienated most of the other physicians in the city, and he finally broke angrily with Baird himself in 1886, when the two quarreled over the younger man's educational credentials.

After the MSWV denied Garrison's application for membership, he had countered in kind. From his post as Wheeling health officer, Garrison began pandering to popular opinion and publicly goading the medical establishment in the press. He fenced incessantly with the city council over financially unrealistic proposals related to sanitation is-sues. Worst of all, in the view of those who saw things as Reeves and Baird had seen them, Garrison did not seem legally qualified to be a member of the state Board of Health: he had not practiced for at least twelve years prior to his appointment, as stipulated in the statute; and he did not possess—by the board's own previously established criteria—a "reputable" medical degree, which the statute had also required for

Dr. George Garrison, the controversial
Board of Health appointee who later
shot Dr. Baird. *Wheeling Daily Register*,
March 8, 1891.

board membership. As all of the principals in this confrontation no doubt remembered, the "reputable degree" clause had been retained in the Board of Health law over the vociferous objections and parliamentary challenges of then-delegate Wilson.[14]

Wilson's calculated insult hit home. All twenty-nine of the other licensed physicians in Wheeling signed a formal petition objecting to Garrison's appointment. Furious delegates at the MSWV's annual convention unanimously endorsed an indignant resolution directed at the hostile governor: "The character of the appointment . . . is not such as the profession have a right to expect, in consideration of the dignity and responsibilities of the office," they protested. Wilson's actions violated both "the letter and the spirit" of the law. Consequently, "The action of the Governor in this appointment merits our unqualified disapproval."[15]

The MSWV censure was widely noted in the national medical press and reprinted in the *Journal of the American Medical Association*. Wheeling newspapers then joined in as well, accusing Governor Wilson of undermining a law that "has given the medical profession of West Virginia so high a standing throughout the country."[16]

Wilson defended his actions by claiming that Garrison met the twelve-year provision by virtue of practicing for six years before passing his exam, though no one offered any evidence of that. Furthermore, the governor claimed that Garrison did hold a degree from a reputable medical school. In a literal sense, the governor was right about that; Garrison possessed an MD from Jefferson Medical College in Philadelphia, one of the most highly regarded and impeccably Regular medical schools in the United States. But the validity of that degree, as the governor well knew, had been hotly and publicly contested.

Back in 1881, Baird had seen to it that his young friend Garrison was among the first physicians to pass the new license examination, but Baird also wanted his protégé to follow up by obtaining the kind of formal degree that the older man considered essential for an aspiring young physician. Accordingly, in 1885, while the two were still on close terms, Baird agreed to finance Garrison's further education. Garrison had gone off to Philadelphia, matriculating at Jefferson Medical College on September 11, 1885. Baird was delighted with that choice. Seven months later, Garrison returned to Wheeling, brandishing an MD degree. Rather than celebration, however, this was cause for confusion and consternation, since Jefferson had long required two years of study for an MD degree.

The records of Jefferson Medical College indicated that the faculty had admitted Garrison directly into the second year of study and voted to award him a degree after one six-month term. This procedure was not unheard of, since Jefferson's catalogue contained an explicit policy of recognizing a first year of study done elsewhere, provided the study had taken place "in some *regular* [emphasis in the original] college, authorized to confer the degree of M.D.," and provided the curriculum followed AMA requirements.[17] Yet Baird knew Garrison had not studied at any medical school prior to 1885. Had Garrison lied to the Jefferson faculty, claiming a first year of study he never took? Had the Jefferson faculty made a mistake? Did the Jefferson faculty waive its own rules and count prior practice in lieu of prior formal study? No one knew for sure what had gone on, but Baird was gravely disappointed and felt betrayed by the young man he had taken under his wing personally, professionally, and financially. This dispute opened the breach between Baird and Garrison that would later have murderous consequences.

Governor Wilson knew very well that Garrison's MD would not have been recognized as fully reputable by the very board to which he was now being appointed. In the eyes of sitting board members, there was little to distinguish the circumstances surrounding Garrison's degree from the circumstances surrounding Arthur Dent's degree; both had been granted by Regular medical schools, but without the required two years of full-time study. Governor Wilson, however, claimed that the standards promulgated by the board in Arthur Dent's case were irrelevant and untenable. The law stipulated only that a candidate hold a degree from a reputable medical college; it contained nothing about the basis upon which the degree was awarded. And Garrison did hold such a degree as far as the governor was concerned, so Garrison would now become a member of the board, protests or no protests. To rub it in, Wilson praised Garrison as a "self-made man," who had been forced to overcome "obstacles, jealousies, and rivalries" to succeed as a doctor—an obvious dig at the MSWV insiders who had recently blocked his application for membership.[18]

Before accepting Garrison's appointment, the remaining members of the board instructed their secretary to write formally to the dean at Jefferson, threatening Jefferson with sanctions unless that school had some satisfactory explanation for Garrison's MD: "The State Board of Health in this State . . . in July 1882, passed certain resolutions (which I am enclosing). If I am correctly informed (which I hope I am not) you will at once see that the Board cannot without stultifying itself treat the Jefferson Medical College other than it treated the Columbus Medical College of Ohio . . . for the same offence." It was true, he added, that Garrison held a license "granted him after an examination by the Board very soon after its organization, when examinations of physicians already practicing were very trivial—such as would not be accepted by the Board today." The board "cannot withdraw our certificate granted to him at that time," concluded the secretary, but its current members intended to uphold the standards established in Arthur Dent's case, even against Jefferson.[19]

The dean at Jefferson quickly replied that no school in the country was more rigorously Regular than Jefferson, and the faculty always demanded two years of scientific study in AMA-mandated subjects, plus a course in dissection, which was also mandatory, before granting a degree. In Garrison's case, however, the Jefferson faculty was under a "mis-conception" regarding the license he had been issued by the West Virginia Board of Health. "When the question of Dr. Garrison's admission to the graduating class was submitted to the faculty, there was opposition to it," but an exception was made on the basis that "Dr. Garrison presented himself with a license which had all the legal authority of a diploma, [so] he was admitted to the senior or graduating class" as if he had already completed another degree elsewhere. Under those misapprehensions, one year was deemed sufficient to ratify his already licensed position by awarding him an MD from Jefferson.

The dean's reply had effectively hoisted the board on its own petard—and the board realized it. To repair damages, the board secretary wrote back to the dean somewhat sheepishly, explaining that he did not mean to imply that the board "played fast and loose in its examinations," but merely to illuminate the special circumstances that led to Garrison's certification:

> In my letter to you I used the word "trivial" as applied to the early
> examinations made by the Board. I desire to make some explanation.
> When the law went into effect (a law not framed to suit the Doctors,
> but a compromise) whilst there were many entitled to certificates
> under the ten year clause, there were many equally deserving who
> had practiced only nine, eight, seven, &c. years and the Board felt it
> their duty to deal leniently with them, as at that time no standard of
> examination had been fixed; so it was left discretionary with different
> members to decide what qualifications should be required, and under
> this rather lax rule many received certificates who would not comply
> with the present requirements of the Board.

That was a remarkably revealing letter, one that Marmaduke Dent would love to have had in his possession when arguing Frank Dent's case before the state supreme court. The board was all but admitting that no real standards existed in 1881, which thus left them free to exercise their extraordinary powers in arbitrary ways. The board had gone easy on the protégé of a prominent MSWV officer, who professed to see the future of medical practice as they did, even though Garrison had practiced barely six years. Yet under the same circumstances, the board had come down hard on both Arthur and Frank Dent, two relatives of the only MSWV president who had challenged the premises of Reevesian rigidity, even though both Arthur and Frank had been practicing longer than George Garrison.

The Jefferson dean thanked the board for its clarification and assured its members that the Garrison case was the only one of its sort he had ever encountered. Since the situation was now clarified and a mistake of that sort would never be repeated, both sides could henceforth resume their positions of mutual trust and reliability. With that, the matter was dropped. But the board could no longer block Garrison's appointment, so at least in the short run, the governor had won. Accordingly, Garrison presented his credentials and joined the board for its July meeting in 1887. Though he promptly tested the waters by proposing a license for a doctor styling himself "the White Pilgrim on Wheels," Garrison soon backed down and went along with the majority on most issues.[20] Nor did Governor Wilson attempt further sabotage. He may have gained a symbolic victory, but he did not want another public showdown with the Regular medical establishment, which was now firmly institutionalized and steadily gaining control of its own future. The board, in turn, resumed its previous courses of action with renewed confidence.

Even though the newest member of the board had passed a "rather lax" license exam himself, the other members steadily and systematically continued to increase the rigor of their license exams, primarily by eliminating individual discretion on the part of the examiners. After 1886, each member of the board began setting tough questions in his own field. The results were predictable—and exactly what Reeves had envisioned. The vast majority of physicians gaining licenses held formal MD degrees. In 1886, for example, the board granted 61 licenses: 53 to graduates of medical colleges deemed reputable; 5 to ten-year practitioners; and just 3 to non-graduates who passed a separate exam. In 1887 the board granted 77 licenses: 62 to graduates of colleges deemed reputable; 14 to ten-year practitioners (the last substantial wave in that category); and only 1 to the single applicant who passed a separate exam that year.[21] At the same time, the board also increased its scrutiny of the degrees being submitted. The American Medical College of St. Louis and the Baltimore Medical College, for example, joined the American Medical Eclectic College of Cincinnati on the disreputable list. By 1890, the board would be recommending, though not yet officially requiring, a three-year curriculum at medical colleges in order to qualify as reputable, and the list of disreputable degrees rapidly grew longer.

While the Dents' appeal to the United States Supreme Court was pending, the West Virginia Supreme Court gave the Board of Health Act yet another endorsement. On September 15, 1888, by unanimous vote in a case known as *State v. Ragland*, the four justices sustained the conviction of a man arrested under Section 20 of the law. Inserted as part of the 1882 revisions of the law, Section 20 was designed to regulate itinerant physicians and drug peddlers through a system of registrations, taxes, and fines. The supreme court justices openly criticized that clause as "a very crudely drawn statute. Its provisions are by no means clear." Even so, they decided not to nullify the entire law because, as they put it, "we think we understand what was intended" in Section 20. And under their interpretation of what was intended, J. B. Ragland's sale of "Ragland's Lightening [*sic*] Relief" violated the law. Though Ragland claimed to be a salesman rather than a physician, the court ruled that he was purveying his product as medicine, and that constituted illegal practice. If the supreme court was unwilling to reject "a very crudely drawn statute" whose "provisions are by no means clear," then the Board of Health had no more to fear from the state courts.[22]

Thus, by September 1888, the West Virginia Board of Health, carrying on the professional project Reeves had initiated over twenty years before, appeared to be in a stronger position than ever before. The board's authority had twice been sustained by their own state supreme court: first and most crucially in the *Dent* case, and again in the *Ragland* case. The board had also weathered the challenge of a hostile governor. And the board was successfully using the power of the law to substantially transform the practice of medicine in West Virginia. Still, the board's future remained in jeopardy. Frank Dent's appeal to the United States Supreme Court was scheduled to be heard in December 1888.

CHAPTER TEN

The United States Supreme Court

The clerk of the United States Supreme Court officially filed Frank Dent's appeal on October 13, 1885, after Frank and his father posted a $300 bond to offset West Virginia's expenses if Frank's appeal failed. Both sides had then settled in to wait while the high court dealt with all the cases filed before theirs. Because the court was obliged at that time to hear all appeals in which plaintiffs sought relief from a state ruling on the basis of a national law or the federal Constitution, the court's docket was crowded. Congress would eventually pass a law in 1891 that offered the Supreme Court some relief by allowing the justices to make final decisions separately in their respective circuits if they did not deem the issues worthy of consideration by the court as a whole.[1] But that law came too late to expedite Frank's case. Indeed, the initial estimate of a three-year wait proved to be slightly optimistic. *Dent v. West Virginia* was finally scheduled for oral argument before the full United States Supreme Court on December 11, 1888, three years and two months after it was filed, and more than four years after Judge Green rendered the decision of the West Virginia Supreme Court.

In the late 1880s the United States Supreme Court held its sessions in the old senate chamber of the Capitol building in Washington, DC; the justices did not move to the neoclassical building that now houses the court until 1935. Chief Justice Melville Fuller presided over the presentation of oral arguments in the *Dent* case, though he was the newest member of the court, having been appointed chief justice by President Grover Cleveland just five months earlier. A battle-tested warhorse of the national Democratic Party, Fuller would remain chief justice for twenty-two years. The eight associate justices sitting with him on the court in December 1888, in order of seniority, were Samuel F. Miller, Stephen J. Field, Joseph P. Bradley, John M. Harlan, T. Stanley Matthews, Horace Gray, Samuel Blatchford, and L. C. Q. Lamar II.

Several of those associate justices had prior personal experiences with medical practice that went well beyond being a patient. Justice

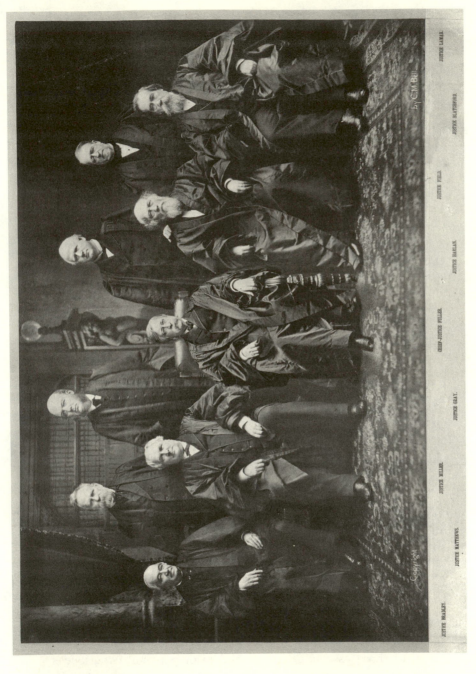

The United States Supreme Court justices in December 1888, when they heard the *Dent* case. C. M. Bell, Collection of the Supreme Court of the United States.

Samuel F. Miller held a formal MD degree from Transylvania University in Kentucky, which taught Regular medicine. He had tried for nearly a decade to make his living as an orthodox Regular physician, purging and bleeding, but eventually abandoned the field in the late 1840s out of frustration over the ineffectiveness of the therapies available to him and the low socioeconomic status afforded physicians. Only then had he turned to law and politics.[2] Justice Samuel Blatchford's uncle Thomas W. Blatchford had been one the most prominent physicians in New York City when the justice was a rising young lawyer there. And Uncle Thomas was well known both to his nephew and to the general public as one of the earliest and most outspoken proponents of forcing non-Regulars from the medical marketplace and upgrading the practice of medicine along AMA lines.[3] Justice Stephen J. Field had once been pressed into service as an ad hoc doctor himself when cholera struck his traveling party while crossing Panama in 1849 on their way to California. He always believed his emergency medical services had saved the life of a fellow traveler.[4]

As the petitioner's attorney, Marmaduke Dent addressed the court first. After submitting the language of the law under which Frank was convicted, Marmaduke reiterated his belief that the statute "was contrary to the XIV Amendment to the Constitution of the United States." Under the writ of error Marmaduke had obtained, that contention was now the sole basis upon which he could challenge Frank's conviction, so his brief before the U.S. Supreme Court would necessarily be much shorter and more pointed than his extended oration before the West Virginia Supreme Court had been. Both the Preston County Circuit Court and the West Virginia Supreme Court, he argued, had failed to recognize—and had refused to apply—the fundamental constitutional principle embodied in the Fourteenth Amendment that "no State shall deprive any person of life, liberty or property without due process of law."[5]

To support his contention, Marmaduke first reestablished the theory that an occupational practice was regarded as a form of property in the eyes of the law; an established occupational practice had the status of an "estate." In support of that point, he quoted classic lines from Blackstone on the expansive meaning of life, liberty, and property, and he cited the Supreme Court's own decisions in the same two ex-Confederate professional practice cases he had cited back in Wheeling: *Cummins v. Missouri* and *Ex parte Garland*. The decision in the *Cummins* case, in particular, was useful to him in the current context, since he could quote Justice Field in defense of the concepts he was trying to establish: that an occupational practice was essentially no different from a house; and state governments did not have the right to strip citizens of their property, real or otherwise, without due process. "For the State to enact a law forbidding a man the enjoyment of his own house without the consent of an arbitrary board of examiners," Dent argued, "is no more unjust than to provide that a man shall not enjoy the benefits of an established practice without a like consent."

Borrowing a line from his earlier briefs, Marmaduke then characterized the edicts of arbitrary boards as "rather ministerial than judicial and wholly different from that which is meant by due process of law." For property in any form to be forfeited, American justice required a procedurally upright finding of wrongdoing—usually by a jury of peers—and a demonstrably reasonable justification for the forfeiture. The license law had neither one. An appointed board could make rulings on whatever criteria they wished, without justifying either the criteria or the ruling. Moreover, any law being invoked to deprive a citizen of property would also have to apply equally to all persons similarly situated. In this case, that would logically mean all people who had been practicing medicine for any length of time prior to passage of the law, not just those who had practiced fewer than ten years. Yet without any stated justification whatsoever, West Virginia lawmakers had inexplicably applied the new requirements to some previously practicing physicians but not to others.

As a further insult to due process, Marmaduke argued, Frank had been "presumed guilty" under the law he was now contesting. The board, after all, did not have to demonstrate that Frank had done something wrong or something harmful in his actual practice of medicine, either before or after passage of the law; instead, Frank was challenged to "prove his innocence before a tribunal authorized to disregard his proof." Put differently, Frank could not prove his innocence by demonstrating that his patients did as well or better than those of any other physician in the state, since patient outcomes were not among the stipulated statutory criteria for licensing.

Marmaduke also challenged the supposed recourse built into the West Virginia law, Frank's right to retain his practice by passing an exam set by the board. As a practical reality, Marmaduke claimed, that was no recourse at all. It was the functional equivalent, he quipped, of "taking a man's house and telling him he can have it back provided he can get the consent of a board" that took it in the first place, a board that was "clothed with full authority to [continue to] withhold it on the slightest pretext." Furthermore, no objective criteria existed against which to measure the board's completely subjective assessment of what distinguished a passing performance from a failing performance. And even if the board members agreed upon appropriate standards, the standards could not be consistently applied when every candidate was examined separately and every examination was different.

The state legislature, according to Marmaduke's appeal, had thus acted unconstitutionally by authorizing the Board of Health to do something the lawmakers themselves could not have done: summarily deny Frank the continued benefits of "a lucrative practice," which he had built up in a perfectly legal manner over many years with the complete trust and support of his patients and his community. Denying him the right to continue in that occupation was an abuse of state power, as well as an act of arbitrary and discriminatory favoritism by a governmental agency wielding the force of criminal sanctions—almost exactly what the Fourteenth Amendment had been designed to prevent.

Finally, to underline the importance of this appeal as a truly significant test case, Marmaduke concluded his brief by reminding the justices that "there has been no decision touching the questions here raised since the promulgation of the XIV Amendment." Consequently, the Supreme Court now had a responsibility to show the nation that the Fourteenth Amendment protected all citizens in every state from arbitrary acts of any kind—not just racially motivated acts. Special interests at the state level should not be allowed to undermine the great constitutional principle of "due process of law." Whenever questionable legislation at the state level violated that overarching constitutional principle for whatever specious reasons, the United States government itself needed to intervene to protect the victims of that legislation. With that he rested his case.

Defending West Virginia, and hence the Board of Health, was West Virginia's popular and well-regarded attorney general, Alfred Caldwell Jr. Like all the other major state officers of that period, Caldwell was a Democrat. He had been elected in 1884 to succeed Cornelius Watts, who had been Marmaduke's adversary four years earlier, when this dispute was before the West Virginia Supreme Court. Reelected in 1888, Caldwell had been remarkably successful at the state level in defending the validity of several controversial regulatory laws passed in the mid-1880s, especially some that were designed to curtail the local power of railroad corporations. Like Marmaduke, Caldwell also rested his case upon constitutional grounds. But Caldwell shifted the focus away from individual property rights under the Constitution's Fourteenth Amendment to West Virginia's police powers under the Constitution's Tenth Amendment, legal terrain he knew well.

Acknowledging that police powers were inherently difficult to define, Caldwell quoted Blackstone, Kent, and others to establish the "well settled principle of the common law that to prevent great public calamity, such as spread of fire, the ravages of a pestilence, &c., private property might be lawfully taken or destroyed for the relief or safety of the public."[6] In short, state legislatures could exercise their police powers over almost anything within their jurisdictional purview if they entertained a good faith belief that the general well-being was at stake. That, in his view, was what West Virginia's duly elected lawmakers did—after appropriate debate and discussion—when they passed the law now at issue.

With that premise in mind, Caldwell next established that policies related to health and medicine did fall clearly under the jurisdictional purview of state legislators. The United States Constitution delegated "preservation of the public health, in the broadest signification of the words, . . . within the police powers of State Legislatures." The exercise of those powers had been repeatedly sustained, both before and after passage of the Fourteenth Amendment. Hence, West Virginia lawmakers had a perfect right—a duty even—to implement policies designed to advance the public health if they saw ways to do so.

When passing the law at issue, according to Caldwell, members of the legislature had reasoned that "the practice of their profession by physicians and surgeons has a most intimate relation to life, limb, health and the well being of the people." Since medical assistance was often required on an emergency basis, with "no opportunity" for careful selection of attendants, the state needed to make sure all physicians possessed "the proper degree of skill and learning." For that and many other reasons, Caldwell continued, "it is unquestionably within the power of State legislatures to shield the people, under the exercise of police power, against the manifold evils resulting from the practicing of the beneficent calling of physician and surgeon by empirics, quacks and charlatans."

In defense of those police powers, Caldwell cited state supreme court decisions in Nebraska, Minnesota, Illinois, Alabama, New York, and Pennsylvania, all of which sustained the validity of various laws restricting property rights in the name of public health. Those state decisions had all been ratified, in turn, by the United States Supreme Court in various other rulings. If the Fourteenth Amendment was interpreted as Marmaduke Dent wished it to be, "the most salutary results from police statutes and regulations would be prevented. For instance," Caldwell continued, citing the Supreme Court's well-known 1873 decision in the Louisiana *Slaughter-House Cases*, "the slaughtering of stock in improper places in New Orleans could never have been stopped."

Due process was thus not an absolute concept. Legislators had the right to halt "the conduct of business detrimental to the public health," even if the conduct of that business had been legal prior to legislative recognition of its detrimental effects. That was the nature of police power. "In most cases its exercise causes loss or injury to the property of some individual engaged in a prohibited occupation or coming in some way within the prohibition of the law, but as has been shown by the authorities before cited, such loss or injury must be borne for the benefit of the public and cannot render the legislative enactment void by reason of any constitutional limitation."

Caldwell conceded that *Cummins* and *Garland*, the two postwar decisions in favor of ex-Confederate professionals, showed "that the estate which one may acquire in a profession is property." But those two decisions had "no effect upon the right of the Legislature of West Virginia to exercise its police power" over such property. Moreover, the Board of Health law was less arbitrary than the Dents made it out to be. Even though the statute would have been perfectly legal without the examination provision, that provision offered to people in Frank's situation an opportunity to prove themselves. But Frank never "tendered himself" to the examiners. Consequently, the board was perfectly correct not "to invest [him] with the important trusts of a physician and surgeon" when Frank himself seemed so "distrustful of his own professional knowledge and skill."

Caldwell then added another, somewhat surprising rationale for the Board of Health law: "The enactment is . . . praiseworthy," he claimed, not only for protection of the public health, but also "for the advancement of the dignity and standing of [the medical] profession." To defend that premise, he asserted categorically that "the rule of law that physicians and surgeons must be learned in their profession and skilled practitioners is of great antiquity." And as his predecessor had done before the state supreme court, he cited Cooley to defend the legislature's legitimate role in advancing that goal. It made no difference whether outsiders thought the public interest would be advanced by upgrading the dignity and standing of the medical profession; it mattered only that the legislature thought so. Whether such a goal was "reasonable or unreasonable, wise or unwise, is a legislative and not a judicial question, and does not affect the matter now under consideration."

Fortunately in this case, Caldwell continued, there could be no doubt that both the law and its implementation were, happily, not only reasonable but wise. Instead of finessing the fact that the board members were pursuing an overtly Regular agenda, the attorney general seized the offensive by praising them for doing so. "[T]he Code of Medical Ethics, adopted by the American Medical Association," he told the high court, contained the "loftiest, purest sentiment" of medical professionalism ever articulated, and the goals of the AMA, he somewhat blithely asserted, "meet with the unqualified approbation of the best minds in the medical profession." For separate emphasis, he singled out the AMA's resolution that "a regular medical education furnishes the only presumptive evidence of professional abilities and acquirements, and ought to be the only acknowledged right of an individual to the exercise and honors of the profession."

Caldwell summarized this line of argument with still another quote from the AMA Code, one that dealt not with the obligations of the medical profession toward the public, but with the obligations of the public toward the medical profession: "The public ought . . . to entertain a just appreciation of medical qualifications; to make a proper discrimination between true science and the assumption of ignorance and empiricism, and to afford every encouragement and facility for the acquisition of medical education." Since members of the public were not, for whatever reasons, behaving that way on their own, Caldwell concluded, "the West Virginia act concerning the public health is a lawful and constitutional attempt to discharge, in a measure, these obligations of the public to physicians." With almost astonishing boldness, given the venue, Caldwell was essentially inverting standard policy assumptions. The general public—acting through elected representatives—had a responsibility to support and to elevate the practices associated with Regular medicine, even though Regular physicians had no obligation to prove that their practices were actually serving the public health more effectively than the practices, in this case, of a successful and experienced physician who held an Eclectic degree.

That was a remarkable conclusion. Caldwell could hardly have put more plainly the fact that a majority of West Virginia's legislators had been persuaded to cast their lots with the goals and premises articulated by the Medical Society of West Virginia. West Virginia lawmakers, in this view, had realized that the AMA Regulars would control their Board of Health and they were nonetheless comfortable clothing the board with the powers of the state. Those AMA Regulars had been authorized to use their judgment about what actions might best serve the public health, even if that meant exercising criminal sanctions to eliminate practitioners whom they did not wish to license. While the Dents considered the legislative delegation of that authority both unconstitutional and without objective justification, Caldwell stood before the court as a convert to the "true church" of Reevesian medicine. He defended the policy not only as legal but also as a positive step forward. Still, those conflicting assessments aside, the key question was whether the Supreme Court would see the board's actions against Frank Dent as a violation of his due process rights under the Fourteenth Amendment.

On January 14, 1889, five weeks after hearing the case, the Supreme Court announced its decision. When the Dents learned that Justice Field had been selected to express the court's position, they must have been heartened.[7] Some sixteen years earlier, in the *Slaughter-House Cases* that Caldwell cited so approvingly, Justice Miller had carried a 5-4 majority in favor of upholding an intensely controversial exercise of public health police power in Louisiana. But that ruling had been confusing and contested from the outset. Moreover, only three of the justices involved in that case—Miller, Bradley, and Field—were still on the court; and neither Field nor Bradley had agreed with Miller. Quite the reverse: Bradley had endorsed what became an influential dissent written by Field and then added his own view that any statute forbidding "a large class of citizens from . . . following a lawful employment previously adopted" would be "onerous, unreasonable, arbitrary and unjust," regardless of its justification as a public health measure. So even back then, Field and Bradley were convinced that the court should be less concerned about sustaining the police powers of the separate states and more vigilant about making sure those powers were not invoked as excuses to violate constitutional guarantees—including due process—now enforceable by the national government under the Fourteenth Amendment. And in the words of Justice Field, state legislatures did not have the power to prevent anyone from pursuing "the ordinary avocations of life."[8]

Even more optimistically, from the point of view of the Dents, during the fifteen years since the *Slaughter-House* decision, the nation's highest tribunal had consistently expressed an abiding skepticism toward the imposition of narrowly targeted governmental regulations. In the time-honored tradition of the founding fathers, the Supreme Court justices of the late 1870s and 1880s still seem to have believed that their first and most important responsibility was the protection of property rights and civil liberties against abuses of power on the part of those who controlled the government. Consequently, they harbored a deep antipathy to statutory rules that seemed to favor special interests, as distinguished from legislation designed to advance a broadly conceived general interest; and they struck out at policies either privileging or constraining one sector or class of society (whether owners or laborers, shippers or carriers) over other sectors or classes.

As a practical matter, many of their decisions had the effect of constricting the legal leverage of laborers and minorities, since the justices frequently blocked efforts to control perceived excesses of industrialization or advance the social and economic positions of suppressed groups. For that reason, an earlier generation of legal scholars dismissed the Supreme Court justices of this period as little more than shameful enablers of capitalist exploitation and minority oppression. But modern legal historians largely agree that the justices were, at least in their own minds, continuing to defend what they regarded as fundamental American freedoms by "pursu[ing] a power-limiting strategy, restricting government (whether state or federal) to carefully adumbrated spheres of activity."[9]

Among the "carefully adumbrated spheres of activity" the Supreme Court justices had generally conceded to state lawmakers during the years between the *Slaughter-House* ruling and Frank Dent's appeal was the exercise of constitutionally mandated police powers that concerned matters of public health. If the welfare of the public was clearly at risk, the justices were willing to accept statutory policies that sacrificed the interests of a few for the general welfare of all, as long as the policies were applied equally to everyone affected by them. But the court adamantly insisted that lawmakers not use public health merely as a pretext or a plausible-sounding rationale for achieving other—less defensible—

ends. Policies put forward in the name of public health and safety had to be self-evidently reasonable on their face, broadly applicable, and clearly justifiable for the common good.

Justice Field himself had stated those conditions clearly in 1884 when concurring with a unanimous decision in the case of *Butchers' Union Slaughter-house and Live-stock Landing Co. v. Crescent City Live-stock Landing and Slaughter-house Co.*, another Louisiana case involving the validity of disputed regulations that were enacted as public health and sanitation measures. Lawmakers at all levels, Field had averred, were constitutionally forbidden from exercising their health-related police powers in ways that either deprived anyone of "the property which every man has in his own labor" or interfered with the inalienable right that all Americans enjoyed "to pursue any lawful business or vocation."[10] In a separate concurring opinion in that same case, Justice Bradley went even further: "Monopolies are the bane of our body politic at the present day. In the eager pursuit of gain they are sought in every direction. . . . If by [state] legislative enactment they can be carried into the common avocations and callings of life, so as to cut off the right of the citizen to choose his avocation," the Supreme Court intended to invoke the national Constitution to nullify them.[11] Those lines might have come directly from Marmaduke's earlier briefs.

Even as Frank's appeal was pending, the justices had further emphasized their determination to prevent abuses of public health police power. Evoking health and safety, the San Francisco City Council had passed laundry regulations that would effectively put the city's Chinese-owned laundries out of business. The Chinese challenged the law, and their challenge reached the Supreme Court in the case of *Yick Wo v. Hopkins, Sheriff of San Francisco*. In his 1886 decision, again writing for a unanimous court, Justice Field ruled the regulations unconstitutional, not because the city council lacked the power to pass public health and safety regulations (in fact, the court had recently approved other San Francisco laundry regulations), but because the law adversely affected a specific sector of the public—the Chinese—without compelling evidence that the targeted businesses threatened the public health and safety of anyone in the city.[12] In that case, Field evoked the equal protection clause of the Fourteenth Amendment rather than the due process clause, but the case was nonetheless promising from the Dents' point of view, since the statutes struck down had attempted to restrict occupational freedoms in the name of public health.[13]

The Dents' hopes must have soared even higher when they heard the first point in Field's decision regarding their own case: "It is undoubtedly the right of every citizen," he proclaimed, "to follow any lawful calling, business, or profession he may choose, subject only to such restrictions as are imposed upon all persons of like age, sex and condition. This right may in many respects be considered as a distinguishing feature of our republican institutions . . . and the estate acquired in them, that is, the right to continue their prosecution . . . cannot be arbitrarily taken away from them." Those opening salvos were entirely consistent with Field's opinions in employment cases over the preceding fifteen years and with the court's repeated defense of the right to pursue "the ordinary callings of life" and the "common occupations." The Dents could not have hoped for a more ringing opening statement.[14]

When Field then turned to the practice of medicine specifically, how-ever, his tone changed abruptly. "Few professions require more careful preparation," he observed, and doctors dealt with "all those subtle and mysterious influences upon which health and life depend." They had to master "the properties of vegetable and mineral substances, . . . the human body in all its complicated parts," and the vagaries of mental health. And since "comparatively few [ordinary citizens] can judge the qualifications of learning and skill" that any given physician possessed, a state would be justified to insist upon a license as evidence of suffi-cient qualification. This was—in his words—"for the protection of soci-ety." Field made no further attempt to reconcile those criteria with his earlier, often repeated, and constitutionally based defenses of occupa-tional freedom.[15] Consequently, from that point on, the practice of med-icine would no longer be treated by the Supreme Court as an avocation that anyone had a lawful right to pursue. The practice of medicine would henceforth be treated instead as a special case—as a governmen-tally recognized profession—and hence as something that states would be permitted to treat differently from "the ordinary callings of life" and "common occupations," by requiring licenses and imposing criminal sanctions.

Field then denied Marmaduke's analogies to *Cummins* and *Garland*. Both of those cases involved laws "designed to deprive parties of their right to continue in their professions for past acts, or . . . sympathies, many of which had no bearing upon their fitness to continue in their professions." In contrast, the West Virginia law, in his view, had nothing to do with medical practices in the past; it was not designed to punish former behaviors but to protect a vulnerable public in the future. He also cited *Yick Wo v. Hopkins*, but only to emphasize the arbitrary nature of the regulations struck down in that case. Here, he asserted, "there is nothing of an arbitrary character in the provisions of the statute in question." The provisions applied equally to anyone who wished to be a physician, and in the court's view, "legislation is not open to the charge of depriving one of his rights without due process of law if it be general in its operation upon the subjects to which it relates and is enforceable in the usual modes established in the administration of government with respect to kindred matters—that is, by process or proceedings adapted to the nature of the case." Moreover, Frank Dent had no claim that the board treated him personally in an arbitrary, ad hominem, or unfair matter, since he "did not submit himself to the examination of the board after it had decided that the diploma he presented was insufficient."

In sharp contrast to most of the arguments he had made in previous occupational and public health cases, where he generally chastised lawmakers for overreaching, Field then offered an uncharacteristically expansive construction of state police powers in those realms. "The power of the state to provide for the general welfare of its people," he declared, "authorizes it to prescribe all such regulations as in its judgment will secure or tend to secure them against the consequences of ignorance and incapacity, as well as of deception and fraud." In the end, therefore, exercise of those powers must necessarily "depend primarily upon the judgment of the state as to their necessity." As long as the regulations adopted "are appropriate to the calling or profession" and realistically attainable, courts should not object. West Virginia legislators obviously considered the approach they took to be the one best suited to optimizing the public's health in their state. Legislators in other states were free to decide that other approaches were better suited to their circumstances, and free to decide that no intervention was necessary in their state. But West Virginia's lawmakers had the right to make the call they did in exercise of their police powers.

Finally, with regard to the key question of due process, Field offered a disarmingly candid observation: "[I]t may be difficult, if not impossible, to give to the terms 'due process of law' a definition which will embrace every permissible exertion of power affecting private rights and exclude such as are forbidden." In other words, due process was whatever the court considered it to be in any given situation. And in this one, a unanimous court was siding with the state legislature and the West Virginia Board of Health. Should the West Virginia Board of Health behave in an "unfair or unjust" manner, Field continued, "we doubt not that a remedy would be found in the courts of the State." But as to the right of the state to pass and enforce the law it did, he left no doubt. The act "was intended to secure such skill and learning in the profession of medicine that the community might trust with confidence those receiving a license under the authority of the State." That rationale trumped Frank Dent's preexisting property rights, even though he would lose the "estate" he had built up in his previous practice. Consequently, the United States Supreme Court formally affirmed the judgment of the West Virginia Supreme Court.

The Dents had lost again—this time without the prospect of further appeal.

PART FIVE

Implications

CHAPTER ELEVEN

American Medical Practice after *Dent*

U p close, the story behind the 1889 *Dent* decision might appear to be little more than a nasty feud between a cadre of medical professionalizers, on the one hand, and the extended Dent family of traditional physicians, on the other—a feud blown out of proportion by fiercely unbending egos on both sides. Viewed that way, the significance of the story would remain principally local and specific, intriguing primarily to people interested in the history of medical practice in West Virginia. Viewed from a longer perspective, however, that feud far transcended the idiosyncratic individuals involved in it. The two sides were fighting over nothing less than the future status of the medical profession in the United States, and their battle forced the nation's Supreme Court to rule on how the relatively recent Fourteenth Amendment would impact occupational and professional regulation by the states. So there are many reasons to step back and reflect upon some of the major implications of that West Virginia feud and the long-term effects of the Supreme Court decision it brought about.

Perhaps most obviously, a long-term perspective serves to remind contemporary Americans of something medical historians have long known well: the practice of medicine in the United States has not always been a licensed profession. Quite the contrary: through 1880, medical practice was neither licensed (in a meaningful or mandatory manner) nor a profession (in a legally empowered and formally recognized sense) anywhere in the United States. Because medical licensing took place on a state-by-state basis over the four decades following 1880, and because each state had its own unique mix of personalities, traditions, socioeconomic realities, and political dynamics, the process by which the practice of medicine became a licensed profession in the United States resembled the making of a mosaic. Though basic patterns emerged, each tile was unique. Consequently, what happened in West Virginia cannot be taken as an exact model of what would happen everywhere else, especially given the exceptional amount of control achieved by West Virginia's most elite AMA Regulars from the outset. Even so, what happened in West Virginia was not some aberrant outlier either. The West Virginia experience established enduring patterns repeated elsewhere throughout the country, and all of those other state-by-state battles over medical licensing were fought out against the backdrop of the *Dent* decision.

The Dent story also reveals without doubt that West Virginia's medical license law was not a regulatory reform demanded by the general public. Nor were its underlying assumptions—much less its specific terms—somehow the products of a seemingly inevitable process of modernization or the inherent consequences of an emerging public consensus. Instead, it was a consciously engineered policy, drafted and passed through the concerted efforts of a specific subset of physicians, the elite Regulars associated with the Medical Society of West Virginia. Indeed, those physicians forthrightly acknowledged their lack of support from the general public, and the champions of medical licensing repeatedly expressed their contempt for ordinary people, whom they regarded as incapable of knowing their own best interests. Similarly, the state's lawmakers right through the 1870s, though charged with protecting the public, had never independently considered the need for mandatory medical licenses. Only after the MSWV organized a campaign of its own, only after MSWV candidates gained seats in the legislature, and only after James Reeves and George Baird struck an alliance with James Ferguson and West Virginia's most powerful corporate interests was medical licensing secured. Even then, the champions of licensing had to overcome substantial popular and legislative skepticism.

Both the vast majority of ordinary citizens and the vast majority of sincere physicians of any sort in the 1880s would surely have agreed that state legislators were justified in trying to eliminate out-and-out charlatans. And both groups would almost certainly have agreed as well to additional restrictions that had firm rationales and applied equally to all of them. Since the 1830s, after all, many state legislatures had been passing laws of that sort without serious opposition, such as laws that made inoculating against smallpox, rather than vaccinating against smallpox, a criminal offense because inoculation was demonstrably dangerous and vaccination was fully effective.[1] But the license law passed by West Virginia in 1882 and upheld by the United States Supreme Court in 1889 went well beyond areas of limited agreement to effectively restrict every aspect of medical practice to physicians fully trained in the academic sciences as defined by the AMA. And that principle divided the state's practicing physicians in several different ways.

The vast majority of non-Regulars, of course, actively fought Reeves's Board of Health law. They adamantly opposed the principle of basing all medical practices exclusively on standards established by the AMA. Their opposition was not difficult to understand: they were justifiably afraid of being driven from the medical marketplace. Often overlooked, however, were the divisions opened in the ranks of the Regulars themselves. All of West Virginia's leading Regulars were committed in a general sense to the belief that science held the key to medical progress. And most of them—including MSWV president William Dent—were active members of the AMA and sympathetic to that organization's long-term goals. And yet, Regulars like the Dents—along with the general public—remained reluctant to privilege one approach to healing over others in an era when none of the available approaches could demonstrate superior results for patients; they opposed a narrow definition of science that insisted on the primacy of theoretical knowledge over personal experience and privileged laboratory results over rational and practical inquiry on multiple fronts; and they remained comfortable with the idea that physicians could be trained to diagnose ills and administer the beneficial results of scientific research, whether or not the physicians fully understood the research techniques that produced those beneficial results. A physician, in their view, could have full faith in scientific approaches to medicine without being a fully trained scientist.

The Dent story also undercuts a related argument sometimes made even by medical historians, particularly some who were writing during the first two-thirds of the twentieth century. Their accounts of medical licensing led readers to believe that meaningful license laws finally appeared in the United States when they did because the modern research sciences were dramatically transforming the field. During the same decades that Reeves and Baird were fighting to require all doctors in West Virginia to demonstrate an academic knowledge of those medical sciences, Louis Pasteur in France and Robert Koch in Germany were establishing the germ theory of disease and leading what became known as the bacteriological revolution. Other researchers throughout the Western world—after centuries of limited progress—were using the techniques of modern science to learn unprecedented amounts of information about the way various organs, tissues, and fluids of the human body functioned and malfunctioned. The new research sciences were becoming so essential to the effective and successful practice of medicine, went the argument, that even the venal and benighted politicians who typically sat in nineteenth-century American state legislatures realized they would be acting irresponsibly if they continued to allow the public to be treated by physicians who did not fully understand those sciences.[2]

Perhaps in a general sense, that account had some validity. Certainly the rise of the modern medical sciences coincided in a general sense with the coming of science-based medical licenses, and physicians like Reeves and Baird were fervently committed to the premise that training in academic science had the power to transform medical practice from an ineffective and ordinary occupation into a true profession capable of advancing the public health of everyone in the Republic. Further, that interpretation of medical history meshed comfortably with the self-perception of twentieth-century physicians, all of whom were trained to understand the laboratory sciences and to apply what later came to seem like their self-evident benefits.

Yet the Dent story reveals a grave weakness in this traditional line of historical explanation: during more than a decade of impassioned arguments in favor of science-based licenses, Reeves and his allies never offered data to demonstrate that the holders of science-based MD degrees did better for their patients than physicians who lacked such degrees. In their campaigns to win legal licensing, they never cited a single scientific advance to illustrate why doctors suddenly needed to understand the research sciences. In their defense of scientific education, they never once explained why a laboratory knowledge of chemistry would be necessary to administer future therapies that might emerge from chemical laboratories.

Other historians have subsequently discovered as well that many American physicians—even after the advent of licensing based on scientific education—were slow to embrace what now appears in retrospect to be the obvious importance of bacteriology, much less more minor scientific discoveries.[3] Knowledge of the human body was advancing rapidly, to be sure, but the medical sciences enabling those advances had not yet produced significant therapeutic breakthroughs that ordinary physicians could actually use in their daily practices. Put differently, a given physician might be willing to believe that a given bacterium caused a given disease, but the physician still had no way to combat the disease once a patient contracted it—and treating the disease was what the doctor got paid for. Indeed, medical science made only modest therapeutic advances, especially in internal medicine, prior to the early sulfa drugs in the 1930s and the mass production of penicillin, which was not available to the general public until the 1940s.[4]

Defenders of the West Virginia license law of 1882 had implied correctly that requiring a mastery of the academic sciences as prerequisite for practicing medicine would help ensure that all the state's future physicians would be well-educated people and thus presumably capable of making better judgments than less well educated people might make. In the long run, they argued, that would benefit society as a whole, and it probably did. But they did not, and could not, justify mandatory science-based licenses as a way to guarantee that all physicians would be able to understand and apply the new therapeutic advances then emerging from scientific laboratories, because the license law was passed *before* any significant therapeutic advances were made. And even if medical science did produce significant therapeutic breakthroughs in the future, argued the opponents of the law, ordinary practitioners could be taught how and when to administer the new antidotes without having to know the arcane chemistry and biology that underlay their discovery.

In the long and triumphant speech that Reeves gave to his colleagues in the MSWV shortly after passage of the 1882 Board of Health law, he addressed the relationship between medical science and medical licensing directly. On the one hand, he praised what he called "the alleged discoveries of Pasteur and Koch" as harbingers of medical progress, and he ticked off several advances in human physiology and medical biology that had recently emerged from laboratory experiments. He was, after all, a lifelong devotee of closely focused research, and the legislature had just given those who agreed with him the power to begin repositioning medical practice on a foundation of academic science. On the other hand, he realized that the benefits of scientific medicine lay somewhere out ahead. He was excited to imagine that "hundreds of pains-taking investigators are, no doubt, already at work" in laboratories around the world, but he knew their efforts were not yet producing applicable results. License law or no license law, the effort to upgrade American medical practice thus remained what it had been from the beginning: a faith-based crusade. In the meantime, he continued, "humanity waits with anxious hopes for the solution" to disease problems.

Reeves celebrated with undisguised pride the fact that he and his fellow board members were engaged in a process that would make "scientific relations" the new norm among West Virginia physicians. All members of the MSWV, he believed, should share that sense of pride, since their concerted efforts had persuaded lawmakers to accept the licensing criteria contained in the Board of Health law. Given all the circumstances, passage of the act had been "a happy surprise to the medical profession of the State." Surprise or not, the new law they managed to get through the legislature now gave West Virginia physicians a nationally unique opportunity. They could begin repositioning medical practice on a scientific foundation, even though, as Reeves put it, effective therapeutic knowledge was "still far from resting on a stable basis." Indeed, the most optimistic statement he could express in 1882 was his belief that effective therapeutic knowledge was "hopefully moving in the [right] direction" toward the ultimate dream of "Medical Certainty." A man of unshakable convictions, Reeves rarely used the adverb "hopefully."[5]

Contrary to older assumptions, therefore, at the time that the West Virginia legislature granted Reeves and his Board of Health the power to license doctors on the basis of their scientific knowledge, the lawmakers were not acting to make sure their constituents would be treated by physicians who understood an already existing range of demonstrably effective approaches to medical care. Instead, the lawmakers were accepting what amounted to a public policy wager. The elite Regulars of the MSWV persuaded a legislative majority that the most likely way to improve standards of health care *in the future* was to limit the practice of medicine to physicians who demonstrated a mastery of the academic sciences. What the United States Supreme Court then ratified, in essence, was the right of those lawmakers to make that policy wager if a majority of them believed it to be in the public interest.

Fortunately for American society, the policy wager made in West Virginia in 1882—and elsewhere over the next four decades—eventually paid off in the form of significantly better treatments for a host of maladies. For that reason, most contemporary Americans would almost certainly now agree with the Supreme Court's 1889 decision to sustain West Virginia's Board of Health Act. In a forward-looking effort to advance the general welfare, that law implemented a state-sanctioned licensing process that put the future of medical care into the hands of those physicians most fully committed to educational uplift and scientific advancement. By the middle of the twentieth century, as other states followed suit, that licensing process had not only rationalized a previously chaotic medical marketplace but successfully transformed the practice of medicine into a scientific enterprise. Those had been

Reeves's goals from the outset, and with the advent of antibiotics and the explosion of therapeutic advances during the second half of the twentieth century, the policy wager he and his allies persuaded the legislature to make in 1882 paid off for their descendants. But as the Dents and others warned at the time, the exclusivity of the policy endorsed in Field's decision effectively blocked alternative approaches that might have made many of the same scientific advances in any event, might have produced different sorts of medical successes, and might have developed more flexible standards for training physicians in other—perhaps ultimately less costly—ways.

AMA Regulars throughout the country quickly realized that the Supreme Court had handed them a significant victory not only for the principle of medical licensing in general but also and more importantly for the validity of licensing physicians on the specific and exclusive terms they had been advocating. State legislatures were now constitutionally empowered to establish whatever criteria they could be persuaded to embrace, even if no objective data existed to justify the requirements they mandated, and even if accompanying criminal sanctions deprived other physicians of their practices—their legal "property." As AMA leaders were quick to point out, the court's willingness to accept what took place in West Virginia opened possibilities that their members elsewhere should pursue aggressively.

Barely a week after the *Dent* decision, for example, the *Medical News*, an influential national journal associated with the AMA's goals and viewpoints, hailed the ruling and urged its readers to recognize the opportunities the decision had opened up. Under the headline "States May Regulate Medical Practice," an editorial summarized the issues in the case and then commended the justices for their "weighty and memorable" decision. Calling upon Regulars in other states to follow West Virginia's lead, the editor implored physicians to press their legislators for mandatory licensing standards like those secured by the MSWV. And now that the Supreme Court had denied Frank Dent's appeal for relief, it was especially important to make sure all new license laws included serious criminal sanctions for practitioners unwilling to comply. The *Dent* decision had effectively unsheathed the sword of the criminal law to "protect . . . citizens from pretenders and quacks," so the editor hoped the example of West Virginia's pilot project would soon be enabling AMA Regulars everywhere "to repress ignorant or fraudulent practice."[6] The editor of the *North Carolina Medical Journal* followed up a few months later by publishing an excerpt from Field's *Dent* decision and exhorting his colleagues in the Regular state medical society to commandeer the state's medical examining board. Future appointments should follow the West Virginia model, and only "the best men" should be allowed to serve.[7]

Typical of the Regular offensive mounted in the wake of the *Dent* decision was a widely circulated address delivered in 1894 to the Massachusetts Medical Society by the eminent Harvard professor Reginald Fitz. Fitz was then president of the exclusive Association of American Physicians, a limited-membership organization of the nation's most highly placed medical scientists. Fitz urged AMA Regulars everywhere to push for standards like those in West Virginia, whether the general public wanted them or not. "We are told that the legislation is not wanted, since the people do not ask for it," he observed, and hence "physicians, it is true, as a rule, take the initiative" in securing strict license laws. But that was justified, in his view, because most citizens, even many well-educated ones, were nonetheless too credulous to know their own best interests. Now that the *Dent* decision had laid to rest the argument that restrictive medical license laws were an unconstitutional form of class legislation or monopolistic control, physicians committed to the AMA's concept of science should redouble their efforts to enact laws like West Virginia's in every state.[8]

Like many other Regular publications around the country, the *Atlanta Medical and Surgical Journal* reprinted Fitz's lecture with an editorial of ringing support. "The opponents of medical legislation have been made up of those who practice some peculiar ism in medicine, the representatives of low-class medical colleges, the great army of medical black-legs, bunco-doctors, charmers, etc., with their hired attorneys. This lovely combination has tried to argue that 'medical legislation invades personal liberty; is class legislation; tends to obstruct the progress of therapeutics; is unnecessary and not wanted; and has proven a failure.' Dr. Fitz disposes of all these wild assertions," the editor continued, and the time had come for all states to begin licensing their physicians under a single state board controlled by Regulars.[9] AMA Regulars throughout the nation redoubled their efforts toward that goal over the following decades, and their attorneys began routinely defending medical license laws by citing the *Dent* decision, as they have continued to do to the present day.[10]

The *Dent* decision may have had an even greater short-term impact on the behavior of non-Regulars than it did on that of Regulars—yet another reason why the Dent story was so important in the history of American medical practice. Non-Regulars in West Virginia, of course, lost their battle to block passage of Reeves's Board of Health law, but they would continue fighting its implementation well into the twentieth century. Elsewhere across the nation, non-Regulars had long been uncertain over the virtues of state licensing per se but were strongly united in their opposition to license laws stipulating processes or criteria that would effectively exclude *them*. The West Virginia law was a prime example of what they had consistently—and successfully—opposed since the Jacksonian era. Now, however, non-Regulars in other states faced the existence of a law with terms clearly drafted to drive physicians like them from the medical marketplace, and the Supreme Court had told their West Virginia counterparts that they had no recourse under the Constitution.

Eclectics in particular felt threatened by the rejection of Frank Dent's Eclectic degree as "disreputable." The editor of the *Transactions of the National Eclectic Medical Association* responded by blaming Frank Dent's predicament—as well as his eventual defeat in court—not on medical realities but on an essentially political factor: West Virginia had lacked an organized Eclectic presence at the time the Board of Health law was passed and implemented, and hence Eclectics had no coordinated way of influencing the legislature or defending themselves as a group against arbitrary actions of the board. That contention ignored the fact that the bitter feud between the Dents and the Wheeling elites had been largely intra-Regular. The festering animosities between those two factions had come to a head only after Frank's father, in his presidential address, urged the MSWV to consider a more flexible and inclusive approach to medical science than the one Reeves and his allies had doggedly advanced for a decade, and only after Frank decided to acquire an Eclectic MD on his way home from the Midwest. No members of the Dent family had previously identified themselves as Eclectics; they had all been active Regulars and members of the MSWV. Moreover, the status of Frank Dent's Eclectic degree was not legally at issue before the Supreme Court. The sole question was whether the West Virginia legislature had the constitutional authority to empower a board of health to take the actions against Frank Dent that James Reeves had taken.[11]

Still, the editor had a point. In states where the Eclectics, Homoeopaths, or other non-Regulars were numerous and well organized, they had been holding their own through the 1880s and into the 1890s against what most of them considered to be a nationwide conspiracy of AMA Regulars. The non-Regulars correctly feared that the AMA was determined to push for license laws designed ultimately to drive non-Regulars from the field. Published exhortations to precisely that kind of legislative action appeared frequently and repeatedly in both the *Transactions* of the AMA and the *Journal of the American Medical Association*, exhortations that fed non-Regular fears and strengthened non-Regular defensive resolve. Consequently, various groups of non-Regulars in over twenty states had already struck working alliances with one another either to resist license laws altogether or to insist that any legislation that might get passed in the future include language that explicitly permitted the licensing of Eclectic, Homoeopathic, and other non-Regular physicians in addition to Regulars. Prior to the *Dent* decision, well-organized non-Regulars in seven other states had even persuaded their legislators to create separate licensing boards for their own physicians.[12]

In that gradual manner, according to the editor of the *Transactions of the National Eclectic Medical Association*, the status of medical practice as a whole, both Regular and non-Regular, was being upgraded in a sensible fashion. Rather than defining the practice of medicine in unnecessarily narrow terms, those states were empowering legitimate practitioners of all sorts—Regular and non-Regular—to exclude the out-and-out frauds none of them wanted to sanction. But in West Virginia, he argued, the absence of organized non-Regular opponents had allowed Reeves and the arch-Regulars associated with the MSWV to craft legislation that effectively granted them exclusive control over the criteria for licensing, thereby allowing them to exclude not just those whom no group wanted but anyone without a degree based on the AMA standards that the board members unanimously favored.

The precedent set in West Virginia was thus the Regulars' fondest dream and the non-Regulars' worst nightmare. The overwhelming majority of Eclectics, Homoeopaths, and other non-Regulars believed the United States Supreme Court should never have allowed such a narrow and exclusively Regular statute to stand. In their view, the West Virginia law created a monopoly that was both unjust and unjustifiable. Yet the court sustained the law, and by legitimating Reeves's actions against Frank Dent as constitutionally permissible, the court had not only altered the legal landscape of medical licensing in a general sense but also opened the door for future licenses to be based upon far more restrictive criteria than any previous American statutes had ever imposed.

As a result, from 1889 onward, significant numbers of non-Regulars, especially within the ranks of Eclectics and Homoeopaths, abandoned their previous battles for independent recognition per se and threw their weight instead behind cooperative standards of practice that would allow most of their members to obtain a license and at least some of their medical schools to slip quietly under otherwise Regular-erected legal umbrellas. This merging toward a single dominant standard of medical practice was one of the most striking developments in the history of American medicine between 1890 and 1920, and it took place, by no accident, only after the *Dent* decision of 1889 forced non-Regulars to face the fact that they could be squeezed from the profession altogether if more states went the way of West Virginia.

While the reaction of medical observers to the *Dent* decision was largely predictable, the reaction of many legal scholars was not. Especially in the short run, many practicing attorneys and most constitutional commentators were openly puzzled by Justice Field's opinion. More than a few influential commentators expressed serious concern. The source of their unease was heightened confusion over exactly what "due process" actually meant under the Fourteenth Amendment and hence just how far state lawmakers would be permitted to extend their police powers. Legal analysts found themselves struggling to understand the decision, to explain what appeared to be its apparent inconsistencies, and to reconcile themselves to its possible interpretations.

For fifteen years Supreme Court justices had scrutinized and rejected statutes that looked to them like exercises of legislative power designed to benefit those who controlled or unduly influenced legislative bodies, as distinguished from statutes designed to benefit the entire society. Though the justices had allowed some economic regulations to stand—perhaps more than they are given credit for—they had consistently challenged rules that seemed arbitrary or without compelling justification, and they had consistently protected property. Many of the regulations they approved, including sanitary requirements and liquor restrictions, had been justified by lawmakers as legitimate efforts to improve public health and safety, since that was a realm in which the federal courts had long recognized the police powers of the states. Even then, however, as noted earlier, the justices had set a high burden of proof, especially under the new Fourteenth Amendment, and they rejected laws that appeared either to have ulterior purposes or to lack persuasive rationales.[13]

In the *Dent* decision, however, the justices suddenly appeared to do an about-face. They inexplicably upheld the right of West Virginia's state lawmakers not only to intervene in the general economic marketplace by creating an occupationally specific regulatory board, but also to grant that board exclusive power to determine on an a priori basis— without external consultation or compelling justification—who would and who would not be permitted to engage in that particular occupation. Moreover, they sustained a law that allowed the new regulatory board, armed with the criminal sanctions of the state, to impose its own standards of judgment unilaterally and absolutely to destroy the preexisting legal "property" of citizens who had previously been earning legitimate livelihoods in that occupation. This seemed to undermine the application of Fourteenth Amendment due process rules in ways that many legal commentators considered both puzzling and troubling.

When the *Central Law Journal* of St. Louis, for example, reprinted the *Dent* decision just two months after it appeared, the editor felt compelled to append a comment. In it he summarized where he thought the questions of due process, police powers, and state licensing now stood in the wake of that decision. On the one hand, he was troubled that the Supreme Court had upheld West Virginia's Board of Health law because it appeared to sanction an arbitrary taking of property by the state without due process. That was an especially sensitive issue for readers in his region, since it seemed so clearly inconsistent with the same court's ruling in *Cummins v. Missouri*, where the absence of reasonable due process had led to the nullification of an occupational licensing law, even before passage of the Fourteenth Amendment. On the other hand, he reassured his readers, he was confident that the *Dent* decision was unique, applicable only to a single specific occupation. It should not be interpreted as opening the door for state legislatures to intervene in the marketplace "under pretense of exercising the police power," and lawmakers, he trusted, would still be barred from enacting any regulations "not necessary to the preservation of the health and safety of the community, that will be oppressive and burdensome to the citizen."[14]

Eight months after the *Dent* decision, George Hoadly addressed the annual national meeting of the American Social Science Association (ASSA), which was held that year in Saratoga Springs, New York. This was a major occasion. Hoadly was one of the nation's most distinguished attorneys and legal scholars. He had also served a term as Democratic governor of Ohio in the mid-1880s before returning to private practice. And the ASSA audience contained a large number of highly placed and politically active people who exercised considerable influence as key consultants and policy makers back in their own home states. In his address, entitled "The Constitutional Guarantees of the Right of Property as Affected by Recent Decisions," Hoadly expressed grave concern about the implications of the *Dent* decision and warned that the time had come to guard the Republic's fundamental property rights against legislative encroachment and judicial waffling.

Hoadly's principal points echoed the arguments Marmaduke had made in his briefs: first, state legislators should not be free to enact statutes overriding due process, especially since the Fourteenth Amendment was specifically designed to strengthen federal protection against such actions; second, in *Cummins v. Missouri* and again in *Ex parte Garland*, the Supreme Court had rightly denied post hoc legislative efforts to deprive citizens of legitimate occupational property without due process; and third, the great liberties of the American Republic rested ultimately upon the bedrock of property protection for individuals against the exercise of power on the part of whoever controlled the state. If the *Dent* decision was now meant to signal renewed tolerance for "an indefinite police power, or an inherent right of government to control property devoted to a so-called 'public' use," then the nation's influential citizens— and especially those gathered for the ASSA convention—needed to consider seriously "whether the Bills of Rights do not need rewriting," and to ponder ways of forcing "courts to open their eyes to the violation of such [individual property] rights" in the guise of preserving the public health. Those were strong words, particularly when addressed to an elite audience of proto-progressive reformers, not corporate lawyers.[15]

A year after the *Dent* decision, Thomas M. Cooley, former chief justice of the Michigan Supreme Court and probably the nation's best-known legal commentator, released the sixth edition of his *Treatise on the Constitutional Limitations Which Rest Upon the Legislative Power of the States of the American Union*, certainly the most authoritative and influential constitutional law treatise of its era. Lawyers on both sides during the state and national Dent trials, as well as all of the judges who ruled on the case in West Virginia and in Washington, had cited and discussed passages from earlier editions of that treatise. Because Cooley's legal theories generally opposed legislative interference in the marketplace, he had previously hailed the United States Supreme Court's consistent curtailing of state police powers during the late 1870s and into the 1880s. But the justices' ruling in *Dent* seemed to ignore his views. He retaliated by relegating the decision to two footnotes and by asserting his own interpretation of what the decision meant.

In the first citation, Cooley used the decision to support the premise that the court would not accept police regulations that did not "affect alike all persons similarly situated," despite Marmaduke Dent's contention that the Board of Health Act affected previously practicing physicians unequally. In the second citation, Cooley ignored the substance of the ruling but quoted Field in support of the premise that "the interest acquired in the practice of learned professions, that is, 'the right to continue their prosecution,' is property which cannot be arbitrarily taken away," despite the fact that Marmaduke had lost the case trying to defend precisely that proposition on explicitly constitutional grounds.[16] Those who read only Cooley could not have known how the Supreme Court had ruled. Thirteen years later, the seventh edition of the *Treatise* slightly revised those footnotes but continued to characterize the *Dent* decision in the same two rather disingenuous ways.[17]

In 1902, the *Michigan Law Review* published an extensive, two-part article on "The Privileges and Immunities of Citizens in the Several States." Written by W. J. Meyers, a prominent attorney in Ann Arbor, the article noted that federal courts, even before the *Dent* decision, had recognized that state legislators had some authority to police physicians. While anyone could practice medicine in whatever manner they wished, legislators could impose general regulations that applied across the board to all of them, Regular and non-Regular alike, such as requiring a knowledge of anatomy in order to perform surgery. But Meyers could not understand the inconsistent treatment of prior practice in grandfather clauses like the one in West Virginia's Board of Health law. "None of the cases" that he found after the *Dent* decision "contain[ed] a very satisfactory discussion" of how the courts could uphold the validity of those grandfather clauses if the need to protect the public health and welfare required all physicians to be certified as fully competent on the basis of new criteria. "It must be said," he concluded, "that this holding [i.e. *Dent*] seems somewhat arbitrary" as an interpretation of the Fourteenth Amendment.[18]

As late as 1904 the eminent Ernst Freund, then one of the nation's leading legal scholars and the metaphorical father of administrative law in the United States, clung to the conviction that a person's professional standing—particularly as a physician—was a protected property right, which states should not be allowed to take away from individuals without due process. In his influential text on *The Police Power, Public Policy and Constitutional Rights,* he criticized state legislatures and state medical boards that forced doctors entering their borders to pass license exams when they did not impose a similar restriction on attorneys. Freund still believed that all medical practitioners should be regarded by states as competent and treated accordingly, not denied the right to continue to practice if they moved to a state with a license law.[19]

A year later, in an essay on due process, law professor and educational reformer Isaac Franklin Russell likewise continued to insist "that the right to continue the practice of the learned professions is property which cannot be arbitrarily taken away."[20] Three years after that, Horace W. Fuller, who published a national magazine of legal commentary, was still upset with Field's comments in the *Dent* decision concerning the inherent uncertainty involved in trying to determine what constituted due process. Fuller called instead for clarification and uniform guidelines.[21] Thus, for more than fifteen years, legal authorities remained puzzled by—even openly critical of—the legal logic behind the *Dent* decision. It seemed to fly in the face of the court's previously established tendency to use the Fourteenth Amendment to protect, in the vocabulary of the amendment itself, the "property" of "persons" against arbitrary governmental actions.

Only after the fact did the principal rationale behind the *Dent* decision become clear, and that rationale constitutes the primary reason why the Dent story deserves a prominent place among the most significant turning points in American medical history, perhaps even American social and legal history. Early in his opinion, Justice Stephen J. Field had conceded forthrightly,

> It is undoubtedly the right of every citizen of the United States to follow any lawful calling, business, or profession he may choose, subject only to such restrictions as are imposed upon all persons of like age, sex, and condition. This right may in many respects be considered as a distinguishing feature of our republican institutions. Here, all vocations are open to everyone on like conditions. All may be pursued as sources of livelihood, some requiring years of study and great learning for their successful prosecution. The interest, or, as it is

sometimes termed, the "estate," acquired in them—that is, the right to continue their prosecution—is often of great value to the possessors, and cannot be arbitrarily taken from them, any more than their real or personal property can be thus taken.[22]

That fundamental and unrestricted right to pursue any "lawful calling, business, or profession" had long applied to what the court routinely called in prior decisions all "ordinary avocations."[23] But because medicine involved "subtle and mysterious" practices, in Field's phrasing; because physicians had to make unusually complicated judgments of critical importance; and because lay persons needed protection "against the consequences of ignorance and incapacity," the court formally and explicitly separated medical practice from all the "ordinary avocations" that Americans otherwise had a nearly absolute right to pursue.

By making the crucial distinction between "ordinary avocations," on the one hand, and the practice of medicine, on the other, the Supreme Court had quite literally enabled the creation of a new profession in the eyes of the law: that of state-licensed physicians. West Virginia physicians thus joined lawyers and military officers as members of a legally empowered profession. As in the case of those two occupational categories, states could now grant physicians the nearly absolute right to determine their own membership, evaluate their own performance, and be rewarded for their efforts, as distinguished from the outcome of their efforts. That last distinction was a particularly important one for the new professional category of physicians, since it settled forever a previous uncertainty in American courts about whether the doctor-patient relationship constituted an implied or de facto contract, which would have held physicians accountable for performing cures if they wanted to get paid.[24] State-level lawmakers could also establish whatever administrative rules they thought necessary for the practice of medicine, and they could delegate the determination and enforcement of those rules to whatever agencies they chose. No state, of course, was required to vest physicians with professional powers, but in states where lawmakers chose to do so, citizens who disagreed with the special rules could not look to the Fourteenth Amendment for relief.[25]

The crucial significance of gaining professional recognition in the eyes of the law had been evident all along to Reeves and his allies and was never more obvious than on the eve of the 1881 session of the legislature, the session that would pass the first Board of Health law. Bracing his colleagues for the political battle ahead, Charles Ulrich, a prominent member of the Wheeling medical elite, addressed his MSWV colleagues on the familiar subject of "Elevating the Standard of the Medical Profession." Like many others before him, he began with ritual condemnation of the state's willingness to tolerate "so many ignoramuses turned loose to prey upon the lives and pockets of mankind," and he predictably identified higher educational requirements as the best means of improving medical care. But higher educational standards by themselves, he recognized, were meaningless without legal authority from the state to enforce them. Only "when laws enacted by our legislature are rigidly and universally enforced" would West Virginia physicians finally be in a position to turn away all but the best-educated physicians and thereby create a true profession, which "men will respect . . . and be willing to remunerate . . . according to their deserts."[26] Fittingly, Ulrich, who was already wealthy in any event, later hosted the MSWV's victory party at the end of that session to celebrate passage of the license law he rightly recognized as the single most essential step in transforming medical practice into a legal profession.

Physicians thus achieved professional status in the eyes of the law only after they gained binding authority to deploy *state powers* to establish their own standards, to punish interlopers, to be paid for their efforts regardless of the outcome of those efforts, and to police on their own terms the qualification, number, and performance of their members. In the *Dent* decision, the United States Supreme Court formally ratified the legal authority of state legislatures to convey such powers to specific physicians of the lawmakers' own choosing, even without objectively defensible criteria for doing so, and even—this was a key point—if that subsequently destroyed, without ordinary due process, the "property" of other physicians. At the time it was made, this decision was a departure from the positions consistently taken by American lawmakers and American courts with regard to physicians since the founding of the Republic, so contemporary legal commentators were justifiably confused by it. Moreover, the Supreme Court justices made the *Dent* ruling in the context of an explicit test of the due process clause of the Fourteenth Amendment, even though that clause had seemingly erected more formidable barriers to occupational stratification than anything that existed in American law prior to the Civil War.[27]

Not until the 1920s did legal commentators finally settle on a comfortable way to rationalize the *Dent* decision's conversion of medical practice into a legal profession, which had struck early analysts as a confusing aberration. During the thirty years following the *Dent* decision, lawmakers throughout the United States steadily enacted an ever-growing body of economic and social regulations, most of which, in turn, had been ruled constitutional when challenged in court. Cumulatively, the new policies began to shift the burden of proof away from governmental regulations themselves, which were assumed to be acceptable interventions that advanced the general welfare, and put them instead on the parties adversely affected by the regulations, who were assumed to be defending a special privilege that somehow inhibited the general welfare. Once the legitimacy of governmental intervention seemed more the norm than the exception, the *Dent* decision appeared in retrospect to be less radical than it did when it was delivered—but only in retrospect.

Rodney L. Mott's treatise on due process, which appeared in 1926, epitomized the post hoc reconciliation of the *Dent* decision. In that often cited text, Mott quoted the same line from Field's decision that Cooley had relegated to a footnote more than thirty years before: "legislation is not open to the charge of depriving one of his rights without due process of law if it be general in its operation upon the subjects to which it relates." What Cooley had regarded as a caveat, however, had emerged over the following three decades as a de facto defense for many of the regulatory statutes passed during the so-called Progressive Era of the early twentieth century: as long as restrictions applied across the board, they were deemed not to violate due process. Consequently, by the mid-1920s, commentators interpreted the *Dent* decision not as a puzzling departure or an aberrant exception, but as perhaps the earliest application of a principle that later justices routinely began to apply to regulations of all sorts: a legal doctrine known loosely as the principle of "equal application." Marmaduke would no doubt have regarded that rationalization of the *Dent* decision—which became standard after the mid-1920s—as bitterly ironic, since he had argued precisely the opposite premise: that the West Virginia Board of Health law applied *unequally* to his cousin Frank.[28]

Even looking back from the 1920s, however, the West Virginia Board of Health law still differed from most subsequent regulations in a subtle way that would have profound implications for the development of medical practice in the United States. Reeves and his allies had been fixed from the outset on requiring every physician to obtain a license based on the standards they considered essential to the advancement of medical practice. They succeeded dramatically in securing and defending that goal, which subsequently enabled state lawmakers across the country to advance the practice of medicine to the legal status of a profession. But either intentionally or inadvertently their 1882 statute said nothing whatsoever about issues that might arise *after* a license was granted. It contained no language giving members of the board any authority over physicians once they obtained a license in the first place and no language imposing further responsibilities or forms of oversight as quid pro quo for the exclusive status being granted and protected by the state. In the *Dent* decision, the Supreme Court thus approved for doctors a *form* of licensing that might be called "preparation-only" licensing, as distinguished from "preparation-and-performance" licensing, which had long been the norm for non-professional licenses.[29] Preparation-only licensing, in turn, had at least two corollary consequences for the subsequent development of the newly legalized medical profession in the United States.

One of the consequences of a preparation-only license was the lack of any after-licensing obligations, even to the state. The *Dent* decision thereby enabled a status for physicians in the United States that differed significantly from the status of medical professionals in most other Western nations, where exclusive occupational controls and professional privileges typically came with public oversight of various sorts and with specific responsibilities for the public good. Put differently, under the 1882 West Virginia law—and similar ones enacted by other states after the Supreme Court ratified West Virginia's—American physicians could invoke the power of the state to control access to their profession and to guarantee a special legal status, but they incurred no reciprocal obligations to the general welfare, in whose name the state had justified its bestowal of a privileged position. The West Virginia Board of Health, for example, did not have the right to establish maximum fees in the public interest (though medical societies themselves had tried to fix minimum fees since the founding of the Republic); nor could the board require licensed physicians to serve in any public capacities (such as helping in clinics established by the state to serve the poor).[30]

A second consequence of a preparation-only license was the presumption that, once issued, the license guaranteed competence forever after.

Field himself seems to have been uneasy with that presumption, since he included a pointed comment on the subject in his *Dent* decision:

> [The logic behind] imposing conditions . . . to practise in the first instance may call for further conditions as new modes of treating disease are discovered, or a more thorough acquaintance is obtained of the remedial properties of vegetable and mineral substances or a more accurate knowledge is acquired of the human system and of the agencies by which it is affected. It would not be deemed a matter for serious discussion [i.e. it would be regarded as self-evident] that a knowledge of the new acquisitions of the profession . . . should be required for continuance in its practice.

Field thus sensed correctly that a scientific approach to medicine, which was being advanced as the basic rationale for mandatory licensing in the first place, would inevitably and somewhat paradoxically produce ongoing changes in what constituted best preparation and best practices. But he did not strike West Virginia's law on the grounds that it failed to require physicians to employ emerging best practices in order to retain their licenses. Instead, any such after-license obligations to keep abreast of current developments would remain optional, to be imposed or not imposed on licensed physicians by their separate states or their separate state medical boards, whenever and however they saw fit.

The issue Field anticipated took only fifteen years to surface as a major problem back in West Virginia. John C. Irons, a leading physician and later mayor of Elkins, addressed the 1896 annual meeting of the MSWV with the plaintive question, "Should Not Legislative Enactment Authorize the Revoking of State Medical License?" Irons had joined the MSWV the same year the license law passed, and he had obtained his own license on the basis of an MD degree earned that year from the Louisville College of Medicine. In his view, rigorous enforcement of the state's license law since then had produced a definite "improvement" in medical practice, and he was "greatly encouraged to hope for a higher professional standard in the near future." But it was already clear to him, and to many others, that a disturbingly large number of physicians who initially qualified for a license at the time of their application had subsequently slipped, sometimes to the point of becoming, as Irons put it, "a menace to the profession as well as to the community." Yet the Board of Health had no means of assessing the performance of its licensed physicians and no legal authority to revoke the licenses it had issued. Irons regarded that as deeply troubling for the future of the profession.[31]

West Virginia was not alone. Once the *Dent* decision upheld the constitutional validity of preparation-only medical licenses—without reference to subsequent performance criteria or any oversight on the part of the state—the new medical profession had no incentive to monitor its own performance, much less to expose possible weaknesses. If anything, the incentives were reversed. Because the profession wanted the public to trust the preparation-only license by itself as a prima facie guarantee of best practice—or at least genuinely acceptable practice—the revocation of licenses for poor performance would undermine that trust. Not until the 1970s did any state begin to mandate some form of continuing education in order to retain medical licenses, and even then both the requirements and their enforcement varied greatly from state to state—and still do.[32] No state medical board ever seriously imposed any performance criteria, nor until the late twentieth century even tried systematically to monitor medical outcomes in any rigorous or publicly visible manner.[33]

The disincentive of medical professionals to undermine the prima facie validity of a license combined with the lack of performance criteria and the absence of monitored oversight to produce circumstances in which American physicians would rarely lose their licenses. As professional regulations were tightened during the twentieth century, state boards like that in West Virginia did finally begin to revoke medical licenses. But the number of revocations was tiny. Even during the decade from 1990 to 1999, a century after Irons warned his colleagues about the problem of initially licensed physicians subsequently becoming incompetent, the West Virginia Board of Medicine revoked an average of 1.9 licenses per year, in a state that had an average of roughly five thousand licensed physicians. And West Virginia was in no way exceptional in that regard. Some states went years during the twentieth century without revoking any medical licenses.

Not only was the number of revocations extremely small, but most revocations had little to do with medical performance per se. Instead, the physicians had their licenses revoked as a consequence of actions that would otherwise be criminal in any event, such as drug abuse, sexual misconduct, and insurance fraud. Superficial care, slipshod methods, and habitual reliance on outmoded therapies continued to go unreported and unpunished, just as they had in the immediate wake of the

Dent decision. Even allowing for the average of 1.4 physicians per year who voluntarily surrendered their licenses to the West Virginia Board of Medicine during the decade of the 1990s, professional pruning thus remained minimal. And because of the state-by-state nature of medical licensing, nothing prevented a physician who lost a license in West Virginia—for whatever reason—from moving to another state, qualifying for a license there, and resuming practice.[34]

Although Reeves and his allies had justified medical licensing as a protection for the public against incompetent physicians, the lack of performance criteria and the absence of oversight ironically helped mask a distressingly high, though ultimately indeterminate, rate of medical mistakes through the ensuing century. The rate remained indeterminate because medical professionals were never required to keep records of how well their patients did, much less open their results to public scrutiny. Even in hospitals, outcome data were not available as public information until the late twentieth century, and then only through self-reporting, which likely produced inherently conservative tallies. At the end of the century, the federal government commissioned the Institute for Medicine to look systematically into medical safety. Based on 1999 data, its report estimated that up to 98,000 patients a year were dying as a result of preventable mistakes in hospitals and over a million patients a year were injured.[35] Given the institutional oversight and peer interactions that took place in hospitals, it is reasonable to assume that mistake rates by individual physicians working alone "in the field," as it were, exceeded those in more controlled settings throughout the twentieth century.

Those completely unmonitored mistake rates, in turn, helped sustain, and even to strengthen, the nation's preexisting malpractice system, which had arisen at an earlier time, before physicians were effectively licensed by the states. The *Dent* decision approved a licensing process that involved no official monitoring, no grievance procedures, and no incentive for physicians to police themselves. Since the state that issued the license was thus taking no responsibility for the actions of those it licensed, individual citizens, one at a time, felt they had no choice but to continue seeking monetary redress for poor performance from individual doctors in the courts, just as they had been forced to do before physicians were licensed. That perception helped feed what eventually became a huge, expensive, and by any measure utterly irrational medical malpractice industry in the United States, an industry that has added to the overall costs of medical care while doing almost nothing to improve the overall quality of that care. At least in theory, that vast dysfunctional industry might have been minimized—or at least made more rational—had the Supreme Court rejected the West Virginia Board of Health law on the grounds that it offered near monopoly control over the practice of medicine without incorporating any meaningful public oversight, economic restraints, or forms of ongoing quality control.[36]

Nothing in the *Dent* decision, of course, *required* state lawmakers to follow West Virginia's example of empowering AMA Regulars exclusively. But legislators were henceforth constitutionally free to do so if persuaded—for whatever reasons—that such targeted and exclusive grants of state power would improve the quality of health care in their state. As the prominent legal scholar Hugh Emmett Culbertson summarized the effect of the *Dent* decision nearly a quarter century after it was rendered, any state "legislature may recognize one school [of medicine] without recognizing all, if the recognition be in the exercise of the proper classification and for the public welfare."[37] At least in theory, such grants of exclusive state power could be made to non-Regulars if the non-Regulars could persuade their legislators that they were the ones most capable of advancing health care; and legislators were also free to divide licensing authority among different groups of physicians—recognizing some but not others—if that was what they thought best for their constituents. In the short run, then, control over the practice of medicine in any given state would depend less upon objective measurements of success with patients than upon the ability of various groups of physicians to make credible arguments and to muster political support in their state legislatures.

As a practical matter, the powers subsequently granted during the next three decades to specific groups of physicians by state legislators around the country were seldom as absolute or as exclusive as those granted to the AMA Regulars under the West Virginia law of 1882.[38] And even in West Virginia, the consolidation of Regular AMA authority went less smoothly than the departed Reeves and Baird might have hoped. Never fully reconciled to the narrow and aggressive ways the Board of Health exercised its licensing powers, the opponents of rigidly defined AMA standards rallied a political counterattack of their own in the early 1890s. Riding a wave of Populist anti-monopolism into the state legislature in 1895, they secured temporary relief from the harsh terms of the 1882 statute by amending it to require formal representation from Homoeopaths and Eclectics on the medical examining committees appointed by the Board of Health.[39]

In the long run, however, the implications of the *Dent* decision inexorably favored the consolidation of American medical practice under the direction of the AMA Regulars, even in those states where the Regulars struck compromises with some of their former non-Regular adversaries. The Regulars represented a substantial majority of all physicians in West Virginia and throughout the rest of the nation, and their rivals were divided into separate organizations. By dominating state licensing boards, the AMA Regulars were able to influence medical school curricula, hospital internships, and specialty certifications, along with a host of other related matters. As a result, over the three decades following the *Dent* decision, AMA Regulars succeeded to a remarkable extent in consolidating the nation's new profession state by state, largely on their own terms.[40] In West Virginia specifically, that process culminated in 1921, when the MSWV, with political and financial help from AMA headquarters, secured passage of a state law requiring all future applicants for medical licenses to hold degrees from medical schools formally approved by the AMA's national educational committee.[41]

Even though Regulars wanted the growing number of license laws throughout the nation to be more vigorously enforced than they sometimes were, and even though Regulars never achieved absolute control over the medical marketplace, their overwhelming influence—exercised and coordinated at the national level by the AMA—would have a profound impact on both the medical practices and the economic philosophies that came to dominate the American health care system during the twentieth century.[42] A host of complex factors produced that system, and most of them have been explored in other books. And needless to say, the American health care system remains the subject of intense public debate and shifting government policies today.[43] But in the end, for all of its magnificently positive results, as well as some of its drawbacks, the unique medical system that evolved in the United States during the twentieth century rested on the bedrock foundation of the new professional licenses authorized by the Supreme Court in the 1889 *Dent* decision.

Epilogue

March 8, 1891, the day after George Garrison shot and killed George Baird in downtown Wheeling and then surrendered himself to the sheriff, the *Register*'s bold headline announced "An Awful Tragedy." According to the paper, men and women of the city spoke of nothing else. Everyone recognized the names of the two physicians involved, both of them public figures and prominent doctors for many years. Though Garrison had recently won election as city health officer—in a race against Baird's son—public opinion turned sharply against him as news of the shooting spread. Emotional outrage punctuated conversations. Some reporters feared mob violence and vigilante action against Garrison. The sheriff posted guards around the jail.[1]

The city's leading physicians met the day after the shooting to pay homage to their colleague George Baird, the fallen Medical Society of West Virginia champion who had helped James Reeves maneuver the Board of Health Act through the legislature. The city's Regular physicians had unanimously opposed Garrison's appointment to that state board four years earlier, and they now rallied both to defend Baird's memory and to denounce the renegade Garrison as a murderer. Prominent physicians from all around the state converged on the city to pay their respects. Speakers at the meeting acknowledged the "fearless" absolutism of Baird's commitment to rigidly defined AMA standards, whether inside the MSWV or out in the public forum. They also conceded that Baird "had in his tongue the gift of bitter sarcasm, in the use of which he was not sparing, or careful, particularly so when Dr. Garrison was concerned." But no amount of personal animosity justified what looked to most of them like a cold-blooded execution.[2]

In his statement to the police, Garrison claimed he had shot in self-defense, afraid that Baird was about to make good on often repeated taunts, threats, and vows to see Garrison's career ended for good. The two doctors had originally fallen out after Garrison returned from Philadelphia with his one-year MD degree, which Baird and the Board of

Health considered unacceptable. Feeling personally and professionally betrayed by his former protégé, whom he had earlier helped license by examination, Baird turned upon Garrison in the same unforgiving way that he and Reeves had always treated apostates. After Reeves retired to Tennessee, Baird organized the MSWV's effort to block Governor Wilson's appointment of Garrison to the state Board of Health. Baird was further outraged when Garrison bested Baird's son in the 1889 election for city health officer. After that, Baird began regularly shouting slurs and curses at Garrison whenever the two met, as he had done the morning of the shooting.

Garrison had retaliated against Baird's public abuse by serving his former mentor with an indictment charging failure to report a case of smallpox. Baird denied there had ever been a case to report, publicly belittled Garrison's diagnostic skills, and ignored the charge. Baird then continued his offensive, first by challenging the accuracy of Garrison's death certificates and then by persuading the county council to remove Garrison as physician to prisoners in the jail.[3] The older physician had even physically assaulted the younger physician six months prior to the shooting, when the two happened to meet in city hall. That incident produced a short fisticuffs, followed by rebukes to both parties from the sheriff and token fines for assault. Since then, Garrison pointed out, Baird had "threatened to blow my head off someday."

Following the city hall altercation, Garrison bought a pistol, took target practice, and perfected a quick-draw move. He also started going about the city armed, claiming to be afraid that the vitriolic and unbending Baird was genuinely prepared to kill him. When the two physicians met that morning in front of Prager's paint and wallpaper store, Garrison told the authorities, Baird appeared to make a gesture toward something under his lapel. Claiming to fear for his life, Garrison said, he had reflexively pulled the weapon he carried underneath his own coat and fired.

The city of Wheeling suspended all public business on the day of Baird's funeral. He had, after all, been mayor of the city during the Civil War and was deeply involved in many aspects of its civic, political, charitable, educational, and economic life for nearly four decades. A thousand people lined the streets to pay their respects to the dead doctor, despite a spring snowstorm that turned those streets to muddy slush. The coroner's jury subsequently brought forth a verdict that attributed Baird's death to "two gun-shot wounds fired from a pistol held in the hands of one George I. Garrison," and the state's attorney for Ohio County promptly indicted Garrison for murder. The press correctly predicted that "the case promises to be one of the longest and most important criminal causes ever put on trial" in the circuit court at Wheeling.[4]

Before the trial could get underway, the court had to empanel a jury. After a week of interviewing potential jurors, the presiding judge realized how difficult it would be to seat an impartial jury. Almost everyone in the city had opinions about the case, and many had close medical, business, or personal relationships with either the victim or the defendant. Blackburn B. Dovener, who headed the prosecution team, and William W. Arnett, who headed the defense team, took turns rejecting the few prospective jurors who could claim no bias. During the second week of interviewing, the presiding judge ordered the sheriff to go out to the rural areas of the county and summon potential jurors who were less directly involved with the principals in the case and less likely to have discussed the affair at great length with everyone they met. Finally, after interviewing 354 potential jurors, twelve men were formally seated.[5] The *New York Times*, which was following the case for a national audience, noted that only one of the jurors lived inside the Wheeling city limits.[6]

As the case unfolded in front of packed crowds day after day, the state argued that no citizen had the right to kill another unless truly imperiled himself, and Garrison had not been so imperiled. Indeed, since the bullets hit Baird from behind, the dead man had been turning to tether his horse, not to confront Garrison. Nor would a second shot have been needed if Garrison's reaction was merely reflexive. Finally, prosecutor Dovener, who would later become a Republican congressman, suggested that Garrison had deliberately provoked the confrontation—by aggressively approaching Baird and demanding in this instance that the older doctor take back his customary imprecations—in order to commit the killing under a false pretext. If the younger doctor felt personally and professionally persecuted, he should have sued the older doctor in court, not shot him.

The defense rested its case primarily on the plausibility of Garrison's perceptions that morning, which in turn rested on the larger—and almost cultural—question of provocation: how much insult and verbal abuse can one man in our community be expected to take unilaterally from another, asked the defense attorneys, before perfectly valid emotions begin to cloud normal reactions and perceptions? It might be true

The first Garrison trial jurors visiting the scene of the shooting. *Wheeling Daily Register*, May 12, 1891.

that Garrison shot Baird under ambiguous circumstances, but Garrison was justified in fearing for his own life. Baird's relentless hounding, unceasing insults, actual threats, and avenging behavior over the previous five years had created those ambiguous circumstances, and they finally did him in. Garrison had been the victim of "one of the most systematic, devilish and malignant persecutions that ever a man was called upon to bear."[7] Without actually claiming what might be called an "honor" defense, Arnett, who had been a Confederate colonel, thus tried to persuade the jurors that Garrison's reactions were both understandable and forgivable.

Both sides concluded that the intense animosity between the two physicians had grown directly from the same issue that impelled the Dents to pursue their case to the United States Supreme Court: the issue of medical credentials. Carl Ulrich, an MSWV stalwart in the 1881 and 1882 battles for a license law on AMA terms and host of the victory party for Reeves and Baird after the law passed, noted on the witness stand that Garrison had a quarrelsome personality, to be sure, and that his self-serving political posturing had alienated all the other licensed Regulars in Wheeling. Yet their "original objection [to Garrison] was not that he was not peaceable, but that he was not a graduate. I objected to him on that myself."[8]

After twenty-one days of proceedings, the jury retired to consider the case. Though the press lavished praise on the prosecution's summation, and most observers, including the reporter for the *New York Times*, thought the state had made the stronger case, editorials speculated that the jury would probably be conflicted. To everyone's surprise, however, the jury returned quickly with a unanimous verdict: they found Garrison guilty of second-degree murder. Arnett moved immediately to set the verdict aside and retry the case, claiming to have new evidence that at least three of the jurors had concealed strong bias against Garrison when they were originally questioned and empaneled. The presiding judge allowed Arnett to present affidavits that impugned the professed neutrality of the challenged jurors, and following several more weeks of quibbling and threats to hold the jurors in contempt of court, the presiding judge declared a mistrial. But he refused to release Garrison and ordered that the case be retried in a later session of the circuit court.[9]

Garrison's second trial began a year after his first and proceeded much more expeditiously. A jury was seated in less than a day, and the new presiding judge gave the defense an initial victory by ruling that Garrison could not be retried for any offense greater than the first verdict of second-degree murder. A few new witnesses were heard, including an African American passerby who thought Baird might be yelling at him when he shouted a racial epithet at Garrison that morning. Garrison himself took the stand in an effort to persuade the jurors that he thought Baird was going for a gun. This time the jury stayed out for nearly three days before declaring themselves unable to reach a verdict. They all considered Garrison guilty, but they could not agree on his degree of guilt. The closest they got was a vote of ten in favor of voluntary manslaughter and two in favor of involuntary manslaughter. The judge declared a hung jury and ordered an immediate retrial.[10]

The third trial began a week after the second one ended. The old witnesses went quickly through the major points of their now well-worn testimony. The only new witness, another passerby located by the defense, offered his opinion that Baird's posture and body motions on that fatal morning were plausibly consistent with an attempt to draw a gun from under his coat. Garrison again pleaded his own case for firing in what he perceived to be self-defense. The third jury stayed out for four days before returning with a verdict that declared Garrison guilty of involuntary manslaughter. That result turned out to have been a compromise between seven jurors who favored conviction for second-degree murder and five jurors who favored acquittal. The presiding judge accepted their compromise and sentenced Garrison to seven more months of jail time, to be added to the fifteen months he had already served, and a fine of $250.[11] The "great doctors' murder case" had finally ended—a small and sorry personal consequence of the far larger and historically significant battle among West Virginia's physicians over the future of their profession.

ACKNOWLEDGMENTS

Over decades of reading, researching, and writing about nineteenth-century medicolegal issues and policy history, I occasionally came across a passing reference to the case of *Dent v. West Virginia*. Those references were invariably short—often just a sentence—and they usually made a single point: in that decision, the United States Supreme Court affirmed the apparently obvious authority of state lawmakers to license doctors. By the middle of the twentieth century that authority was so clearly established and so seemingly self-evident that no additional analysis of such a pro forma decision seemed necessary.

To me, however, as someone interested in nineteenth-century medicolegal issues and the origins of American social policies, the formal affirmation of state licensing powers seemed on its face to be a major step in the history of medical practice in the United States. How could it not be, I wondered. And if the constitutional authority to license physicians at the state level was so self-evident, why had the case come before the Supreme Court in the first place? Why had the case arisen when it did? Who were these defendants, what was the nature of their contention, and what lay behind the dispute? Did the decision actually change anything or have any impact? So I set about trying to find out a bit more about this seldom mentioned and rather enigmatic case. I naively assumed that would be a straightforward task, quickly accomplished.

I was wrong. The job of satisfying my curiosity proved to be much more difficult than I imagined it might be. The secondary literature offered little help. Neither the long-standard history of medical licensing, which was published in 1967, nor the relevant volume of the Holmes Devise history of the Supreme Court, which was published in 1993, even mentioned the case. Among monographic studies that did note the decision, I found unsettling inconsistencies, which included contradictory dates, Marmaduke as defendant, and Eclectic medicine as the issue. So I decided to look at the original sources myself. Curiosity became research. The more deeply I dug, the more fascinating the story became, and the effort to satisfy my curiosity expanded into a narrative with implications that reverberated well beyond the specifics of the case itself.

Fortunately for me, I had a great deal of help trying to find out as much as I could about the circumstances surrounding *Dent v. West Virginia*. Historians simply cannot work without well-managed collections of original material; and we cannot use those collections effectively without capable reference librarians to guide us through them. I am enormously grateful to have benefitted from several of both. The West Virginia and Regional History Collection, which is housed in the Charles C. Wise, Jr. Library at West Virginia University, is an exemplary archive, where I found a great deal of rich and well-organized material. Christy Venham, an equally exemplary reference librarian, guided me through that collection. She helped from afar by e-mail and scan and then helped in person during an intensive period of research in Morgantown. She was really excellent. Jessica Eddy of the George R. Farmer, Jr. Law Library at West Virginia University aided my legal searches. I would also like to thank the kindhearted custodian at the Farmer Law Library who accompanied me to the repository of state appellate court records one night so I could copy some key documents unavailable elsewhere.

Elsewhere in West Virginia, I continued to find additional information and helpful people. Betsy Castle, clerk of the circuit court for Preston County, kindly had her staff locate the original ledgers pertaining to Frank Dent's initial conviction, and those ledgers were waiting for me—along with all other legal references they had to the Dent family—when I arrived in Kingwood. When I visited Newburg, Charles Plum, the town's unofficial historian, kindly put his business on hold in order to go through scrapbooks, local histories, and maps. He showed me where the Dent doctors had their homes (now gone), and then later he sent photographs and additional information about them. Vanessa Lough verified documents for me in Wheeling. Ellen Briggs fielded questions and led me to valuable data compiled by the West Virginia Board of Medicine. Rory Perry, clerk of the Supreme Court of Appeals of West Virginia, checked his records for me. Richard Ivan Dent was extremely generous with his detailed knowledge of Dent family genealogy and with his collection of old family photographs. He also put me in touch with other members of the Dent family, who shared what they knew about their forebears. I am deeply grateful to Rick for the help he offered to a complete stranger from across the continent.

Another major portion of the research for this book was conducted during several trips to the New York Academy of Medicine. The Academy's extensive and well-run library contains a complete set of the *Transactions of the Medical Society of West Virginia* for the decades I needed, and I spent many hours with those fragile volumes; the Academy staff, in turn, spent many hours copying relevant sections for me. I am grateful to the NYAM for maintaining the valuable scholarly resources they hold and to Arlene Shaner and the rest of the library staff for accommodating visiting researchers with efficiency and grace. Joan Claro and Eric Robinson at the New York Public Library also helped me find and scan old newspapers that were not readily available elsewhere.

In Boston I found valuable material in the microform collection at Harvard University's Widener Library and in the Center for the History of Medicine at the Harvard Medical School's Countway Library. Alex Herrlein at the Lloyd Library and Museum in Cincinnati checked records of the Eclectic medical schools in that city. David Kessler at the University of California's Bancroft Library in Berkeley, Michael Angelo at Thomas Jefferson University in Philadelphia, and Linda Jordan at the Federation of State Medical Boards were among many others who fielded specific questions that came up during the writing of this book. I am grateful to them and to scores of other generous people in governmental offices, medical organizations, and educational institutions who did the same thing, often without realizing they were adding to this particular project; they were just doing their jobs courteously and well.

Most of the work for this book was done at my home institution, the University of Oregon. I am deeply appreciative of the support that institution has given me. The UO's continued commitment to faculty research in the face of contemporary financial crises has been admirable indeed. At the top of the list of many individuals who helped me at the UO is John Russell, then history reference librarian at the Knight Library. He consistently found ways to answer even the most arcane inquiries. The Interlibrary Loan staff diligently borrowed both obscure books and hard-to-find microfilms; and I am especially grateful to Tamara Vidos for her patient assistance with the latter. Angus Nesbit, Andrea Coffman, and the rest of the reference staff at John E. Jaqua Law Library at the UO helped with legal sources. Christian Boboia prepared all of the illustrations for this book. Kathleen Leue arranged for Liz Podowsky to create a customized map. Two colleagues at the UO, Jeff Ostler and R. James Mooney, kindly vetted sections of the manuscript. I would also like to acknowledge the work of two UO students, Elan Ebeling and Clinton Sandvick. The former wrote a prizewinning honors thesis on the politics of public health in Wheeling, and the latter is finishing a dissertation that examines the variegated mosaic of medical licensing laws that emerged across the nation during this era and into the early twentieth century following the *Dent* decision.

Fellow scholars around the country have generously shared their knowledge, helped shape my thinking, and offered specific help of various kinds. Chief among them is William G. Rothstein of the University of Maryland, Baltimore County. Bill and I have been close friends for more than forty years, and our ongoing discussions of medical history have done more to stimulate my thinking about that field than anything else. Bill also read and discussed various parts of this project as I developed them and then read and discussed the entire manuscript when it was finished. I am deeply grateful for all of that, both long-term and short-term. Michael Les Benedict shared his knowledge of judicial issues, and John S. Haller Jr. shared his knowledge of Eclectic medical colleges. Many others too numerous to mention—especially in the American Association for the History of Medicine and the biennial Policy History conferences—made suggestions, asked key questions, or answered phone inquiries. I hope they recognize a telling point here or a detail there that I would have missed without them. Robert McGlone at the University of Hawaii, Manoa; Georg Schild at the Eberhard Karls Universität Tübingen; and Manfred Berg at the Heidelberg Center for American Studies all offered opportunities to present lectures related to this project, which generated valuable feedback from diverse perspectives.

Jackie Wehmueller, my editor at the Johns Hopkins University Press, has been patient and encouraging from the outset, as well as a perceptive reader of the manuscript and the source of several good suggestions about organization and presentation. She has been a pleasure to work with. I also want to thank her assistant, Sara Cleary, who deftly handled a host of practical issues, and Jeremy Horsefield, who provided an excellent copy edit. And I am enormously grateful to the anonymous readers who agreed to review the manuscript for JHUP. I have published several books, but I never before received such genuinely helpful readers' reports; only they will see what a positive impact they had.

Words cannot express how thankful I am for the most unfailingly supportive person in my life, my wife Betty, to whom this book is dedicated. I am fortunate beyond imagining to be sharing my life with her.

CHAPTER ONE: *Medical Regulation in the United States through the Civil War*

1. In general, see William J. Novak, *The People's Welfare: Law and Regulation in Nineteenth-Century America* (Chapel Hill, 1996); R. Rudy Higgens-Evenson, *The Price of Progress: Public Services, Taxation, and the American Corporate State, 1877 to 1929* (Baltimore, 2003); John Duffy, *A History of Public Health in New York City, 1625–1866*, vol. 1 (Russell Sage Foundation, 1968), and *The Sanitarians: A History of American Public Health* (Urbana, IL, 1990); Barbara Rosenkrantz, *Public Health and the State: Changing Views in Massachusetts, 1842–1936* (1972); Charles E. Rosenberg, *The Care of Strangers: The Rise of America's Hospital System* (Baltimore, 1987); and James C. Mohr, *Doctors and the Law: Medical Jurisprudence in Nineteenth-Century America* (New York, 1993). A great many more closely focused histories attest to numerous health-related policies implemented at the state, county, and local levels throughout the United States.

2. My composite portrait of mainstream medical practice through the Civil War draws primarily on William G. Rothstein, *American Physicians in the 19th Century: From Sects to Science* (Baltimore, 1972); John Harley Warner, *The Therapeutic Perspective: Medical Practice, Knowledge, and Identity in America, 1820–1885* (Boston, 1986); John S. Haller Jr., *American Medicine in Transition 1840–1910* (Champaign, IL, 1981); Joseph F. Kett, *The Formation of the American Medical Profession; the Role of Institutions, 1780–1860* (New Haven, CT, 1968); Paul Starr, *The Social Transformation of American Medicine: The Rise of a Sovereign Profession and the Making of a Vast Industry* (New York, 1982); James H. Cassedy, *Medicine in America: A Short History* (Baltimore, 1991); Morris Vogel and Charles E. Rosenberg, *The Therapeutic Revolution: Essays in the Social History of American Medicine* (Philadelphia, 1979); Charles E. Rosenberg, *Explaining Epidemics and Other Studies in the History of Medicine* (New York, 1992); Richard Harrison Shryock, *Medicine and Society in America, 1660–1860* (New York, 1960); Whitfield J. Bell Jr., *The Colonial Physician and Other Essays* (New York, 1975); Henry B. Shafer, *The American Medical Profession, 1783–1850* (New York, 1936); Elaine G. Breslaw, *Lotions, Potions, Pills, and Magic: Health Care in Early America* (New York, 2012); Ronald L. Numbers, "The Fall and Rise of the American Medical Profession," in Nathan O.

Hatch, ed., *The Professions in American History* (Notre Dame, IN, 1988), 51–72; Ronald L. Numbers and John Harley Warner, "The Maturation of American Medical Science," in Nathan Reingold and Marc Rothenberg, eds., *Scientific Colonialism: A Cross-Cultural Comparison* (Washington, DC, 1987), 191–214; and Edward C. Atwater, "The Medical Profession in a New Society, Rochester, New York (1811–60)," *Bulletin of the History of Medicine*, 67 (May–June 1973), 221–235. With few exceptions, I will not repeat specific citations.

3. Most medical historians regard the first great protest of this sort to be Harvard professor Jacob Bigelow's publication of *A Discourse on Self-Limited Diseases, Delivered before the Massachusetts Medical Society, at Their Annual Meeting, May 27, 1835* (Boston, 1835).

4. For excellent data on changing therapeutic practices, see John Harley Warner, *The Therapeutic Perspective: Medical Practice, Knowledge, and Identity in America, 1820–1885* (Cambridge, MA, 1986), especially 83–185.

5. On the role of medicolegal questions in driving advances in medical knowledge, see James C. Mohr, *Doctors and the Law: Medical Jurisprudence in Nineteenth-Century America* (New York, 1993).

6. The quote is from John Harley Warner, *The Therapeutic Perspective: Medical Practice, Knowledge, and Identity in America, 1820–1885* (Cambridge, MA, 1986), 161. On the rise of medical science, see Charles E. Rosenberg, *No Other Gods: On Science and American Social Thought*, rev. ed. (Baltimore, 1997).

7. Readers interested in the history of irregular sects might want to start with the essays included in Norman Gevitz, ed., *Other Healers: Unorthodox Medicine in America* (Baltimore, 1988); and James C. Whorton, *Nature Cures: A History of Alternative Medicine in America* (New York, 2002), 1–130. For more detailed histories of these three specific sects, see John S. Haller Jr., *The People's Doctors: Samuel Thomson and the American Botanical Movement, 1790–1860* (Carbondale, IL, 2000); John S. Haller Jr., *Kindly Medicine: Physio-Medicalism in America, 1836–1911* (Kent, OH, 1997); Alex Berman and Michael A. Flannery, *America's Botanico-Medical Movements: Vox Populi* (Binghamton, NY, 2001); Susan E. Cayleff, *Wash and Be Healed: The Water-Cure Movement and Women's Health* (Philadelphia, 1987); and Jane B. Donegan, *"Hydropathy: Highway to Health": Women and Water-Cure in Antebellum America* (New York, 1986).

8. Martin Kaufman, *Homeopathy in America: The Rise and Fall of a Medical Heresy* (Baltimore, 1971); John S. Haller Jr., *The History of American Homeopathy: The Academic Years, 1820–1935* (New York, 2005).

9. John S. Haller Jr., *Medical Protestants: The Eclectics in American Medicine, 1825–1939* (Carbondale, IL, 1994).

10. On these early licenses, see Richard Harrison Shryock, *Medical Licensing in America, 1650–1965* (Baltimore, 1967); and William G. Rothstein, *American Physicians in the Nineteenth Century: From Sects to Science* (Baltimore, 1972), 63–84, and Appendix II.

11. Alexander Wilder, *History of Medicine: A Brief Outline of Medical History and Sects of Physicians, from the Earliest Historic Period; with an Extended Account of the New Schools of the Healing Art in the Nineteenth Century, and Especially a History of the American Eclectic Practice of Medicine, Never Before Published* (New Sharon, ME, 1901), documents the legislative counterattacks of the non-Regulars.

12. On the history of the AMA, see Morris Fishbein, *History of the American Medical Association, 1847–1947* (Philadelphia, 1947); and James Burrow, *AMA: Voice of American Medicine* (1963). On the history of medical education, see William F. Norwood, *Medical Education in the United States before the Civil War* (Philadelphia, 1944); William G. Rothstein, *American Medical Schools and the Practice of Medicine: A History* (New York, 1987); Kenneth M. Ludmerer, *Learning to Heal: The Development of American Medical Education* (New York, 1985); and Thomas N. Bonner, *Becoming a Physician: Medical Education in Britain, France, Germany, and the United States, 1750–1945* (New York, 1996).

13. Some jurisdictions offered physicians exemption from militia duty, and most military organizations granted officer rank to their medical attendants. But such concessions were a far cry from legal autonomy. For an overview of shifting cultural and rhetorical concepts of professionalism, see the sophisticated analyses in Bruce A. Kimball, *The "True Professional Ideal" in America: A History* (Cambridge, MA, 1992), especially 1–17 and 303–325. Burton J. Bledstein, *The Culture of Professionalism: The Middle Class and the Development of Higher Education in America* (New York, 1976), explores the ethos of professional aspiration in the United States during the nineteenth century. See also Harold L. Wilensky, "The Professionalization of Everyone?," *American Journal of Sociology*, 70 (September 1964), 137–158. On the history of professionalization generally during this era see Eliot Freidson, *Profession of Medicine: A Study of the Sociology of Applied Knowledge* (New York, 1971), and *Professional Powers: A Study of the Institutionalization of Formal Knowledge* (Chicago, 1989); Andrew Abbott, *The System of Professions: An Essay on the Division of Expert Labor* (Chicago, 1988); Samuel Haber, *Authority and Honor in the American Professions, 1750–1900* (Chicago, 1991); Gerald Geison, ed., *Professions and Professional Ideologies in America* (Chapel Hill, 1983); Thomas L. Haskell, *The Authority of Experts: Studies in History and Theory* (Bloomington, IN, 1984); and Frederick Wirt, "Professionalism and Political Conflict: A Developmental Model," *Journal of Public Policy*, 1 (February 1981), 61–93. My approach, however, treats "professional" not as an essentially cultural, theoretical, or sociological status, but rather as a *legal* concept. Put differently, I am defining a profession not as an occupation with a particular set of a priori characteristics and behaviors, which may or may not entitle the members of that occupation to informally implied powers or elevated social standing, but rather as an occupation statutorily empowered with officially recognized and formally enforceable rights of self-regulation and legal immunities not granted to all other ordinary occupations.

14. Exactly why nineteenth-century therapeutics were perceived to work at all has long been a puzzle for medical historians. Recent thinking has followed a line articulated thirty-five years ago by Charles E. Rosenberg in "The Therapeutic Revolution: Medicine, Meaning, and Social Change in Nineteenth-Century America," *Perspectives in Biology and Medicine*, 20 (1977), 485–506, which addressed the role of ritual in the healing process. Contemporary studies are exploring similar phenomena in twenty-first-century circumstances. See, for example, Erik Vance, "Seeking to Illuminate the Mysterious Placebo Effect," *New York Times* (June 22, 2010); and Michael Spector, "The Power of Nothing," *New Yorker* (December 12, 2011).

15. On the use of the famous mercury-based "blue mass pills," see Gert Brieger, "Therapeutic Conflicts and the American Medical Profession in the 1860's," *Bulletin of the History of Medicine*, 41 (1967), 215–222. Some scholars debate the possibility that President Abraham Lincoln and others may have suffered from mercury poisoning as a result of taking those pills through most of their adult lives. See Norbert Hirschhorn, Robert G. Feldman, and Ian Greaves, "Abraham Lincoln's Blue Pills: Did Our 16th President Suffer from Mercury Poisoning?," *Perspectives in Biology and Medicine*, 44, no. 3 (Summer 2001), 315–332. On the AMA votes to endorse bleeding, see Donald E. Konold, *A History of American Medical Ethics, 1847–1912* (Madison, WI, 1962), 34. On the repeal of older license laws, see William G. Rothstein, *American Physicians in the 19th Century: From Sects to Science* (Baltimore, 1972), Appendix II.

16. The quotation is from Bonnie Ellen Blustein, "'To Increase the Efficiency of the Medical Department': A New Approach to U.S. Civil War Medicine," *Civil War History*, 33, no. 1 (1987), 22–41. As with most aspects of the Civil War, there is a substantial literature on the conflict's physicians, which ranges from multivolume government reports published soon after the war to recently annotated memoirs, biographies, and local histories. Two long-standard monographs remain the most often cited: George W. Adams, *Doctors in Blue: The Medical History of the Union Army in the Civil War* (New York, 1952); and H. H. Cunningham, *Doctors in Gray: The Confederate Medical Service* (Baton Rouge, LA, 1958). For more recent discussions, readers might start with Mary C. Gillett, *The Army Medical Department, 1818–1865* (Washington, DC, 1987), chapters 7–13; James M. McPherson, *Ordeal By Fire: The Civil War and Reconstruction* (New York, 1992), 385–390; and Margaret Humphreys, *Intensely Human: The Health of the Black Soldier in the American Civil War* (Baltimore, 2008).

CHAPTER TWO: *Dr. Reeves and the Founding*

1. Charles H. Ambler and Festus P. Summers, *West Virginia: The Mountain State*, 2nd ed. (Englewood Cliffs, NJ, 1958), 272–297; Richard O. Curry, *A House Divided: A Study of Statehood Politics and the Copperhead Movement in West Virginia* (Pittsburgh, 1964); Otis K. Rice, *West Virginia: A History* (Lexington,

KY, 1985), 154–182; John Alexander Williams, *West Virginia: A Bicentennial History* (New York, 1976), 30–94.

2. On early medical practice in West Virginia specifically, see Charles A. Wingerter, "Development of Medical Practice and Public Health," in James M. Callahan, *Semi-Centennial History of West Virginia* (Wheeling, 1913), 538–551; and Sandra Lee Barney, *Authorized to Heal: Gender, Class, and the Transformation of Medicine in Appalachia, 1880–1930* (Chapel Hill, NC, 2000).

3. James E. Reeves, *The Health and Wealth of the City of Wheeling* (Baltimore, 1871), 122.

4. William G. Rothstein, *American Physicians in the Nineteenth-Century: From Sects to Science* (Baltimore, 1972), Appendix II, 339.

5. Norman F. Kendall, *Mothers Day, A History of Its Founding and Its Founder* (Grafton, WV, 1937).

6. "Necrology: James Edmund Reeves," *Journal of the American Medical Association*, 26 (January 18, 1896), 141–142.

7. [Women's Auxiliary to the West Virginia State Medical Association], *Past Presidents of the West Virginia State Medical Association, 1867–1942* (Charleston, WV, 1942), 28–30; L. D. Wilson, "James Edmund Reeves, M.D.," *Transactions of the West Virginia Medical Society* (Wheeling, 1896), 1319–1327.

8. L. D. Wilson, "James Edmund Reeves, M. D.," *Transactions of the West Virginia Medical Society* (Wheeling, 1896), 1319–1327.

9. James E. Reeves, *The Health and Wealth of the City of Wheeling* (Baltimore, 1871).

10. Samuel Gross, *Autobiography of Samuel Gross: With Sketches of his Contemporaries* (Philadelphia, 1887), 156.

11. *Transactions of the West Virginia Medical Society* (Wheeling, 1868), 3–4, 12.

12. Ibid., 12.

13. On the practice of railroads keeping a substantial number of influential doctors and lawyers on retainer, see Robert S. Gillespie, *The Train Doctors: A Detailed History of Railroad Surgeons* (2006), available at http://railway-surgery.org/HistoryLong.htm, accessed October 4, 2010; Ira M. Rutkow, "Railway Surgery: Traumatology and Managed Health Care in 19th-Century United States," *Archives of Surgery*, 128, no. 4 (April 1993), 458–463; William G. Thomas, *Lawyering for the Railroad: Business, Law, and Power in the New South* (Baton Rouge, 1999); and Barbara Young Welke, *Recasting American Liberty: Gender, Race, Law, and the Railroad Revolution, 1865–1920* (New York, 2001).

14. Gibson Lamb Cranmer, *History of the Upper Ohio Valley*, vol. 1 (Madison, WI, 1890), 573.

15. On Hupp, see J. H. Newton, G. G. Nichols, and A. G. Sprankie, *History of the Pan-Handle* (n.p., 1879), 265.

16. *Transactions of the West Virginia Medical Society* (Wheeling, 1868), 7–8.

17. See, among others, William G. Rothstein, *American Physicians in the 19th Century: From Sects to Science* (Baltimore, 1972); and John Harley Warner, *Against the Spirit of System: The French Impulse in Nineteenth-Century American Medicine* (Baltimore, 1998).

18. [Women's Auxiliary to the West Virginia State Medical Association], *Past Presidents of the West Virginia State Medical Association, 1867–1942* (Charleston, WV, 1942), 1–3.

19. *Transactions of the West Virginia Medical Society* (Wheeling, 1868), 18–29.

CHAPTER THREE: *Building the "True Church"*

1. *Transactions of the Medical Society of the State of West Virginia* (Wheeling, 1871), 229–230, and (Wheeling, 1886), 248–260.

2. *Transactions of the Medical Society of the State of West Virginia* (Wheeling, 1877), 238, and (Wheeling, 1887), 248–260.

3. L. D. Wilson, "James Edmund Reeves, M. D.," *Transactions of the Medical Society of the State of West Virginia* (Wheeling, 1896), 1325.

4. *Transactions of the Medical Society of the State of West Virginia* (Wheeling, 1875), 18.

5. *Transactions of the Medical Society of the State of West Virginia* (Wheeling, 1869), 90.

6. *Transactions of the Medical Society of the State of West Virginia* (Wheeling, 1870), 151.

7. John Harley Warner, "From Specificity to Universalism in Medical Therapeutics: Transformation in the Nineteenth-Century United States," reprinted in Judith Leavitt and Ronald Numbers, eds., *Sickness and Health in America*, 3rd ed. (Madison, WI, 1997), 87–101; Warner, *The Therapeutic Perspective: Medical Practice, Knowledge, and Identity in America, 1820–1885* (Cambridge, MA, 1986); and William G. Rothstein, *American Physicians in the 19th Century: From Sect to Science* (Baltimore, 1972).

8. On the early registration laws, which had almost no impact on the actual practice of medicine, see Richard Harrison Shryock, *Medical Licensing in America, 1650–1965* (Baltimore, 1967); and forthcoming dissertation by Clinton Sandvick, University of Oregon.

9. *Transactions of the Medical Society of the State of West Virginia* (Wheeling, 1872), 340–341.

10. *Transactions of the Medical Society of the State of West Virginia* (Wheeling, 1872), 347, 344.

11. On the rising number of malpractice suits during this period, see James C. Mohr, "American Medical Malpractice Litigation in Historical Perspective," *Journal of the American Medical Association*, 283 (April 5, 2000), 1731–1737;

and Kenneth A. DeVille, *Medical Malpractice in Nineteenth-Century America: Origins and Legacy* (New York, 1990).

12. *Transactions of the Medical Society of the State of West Virginia* (Wheeling, 1874), 526, 543–544; Charles A. Wingerter, "Development of Medical Practice and Public Health," in James Morton Callahan, ed., *Semi-Centennial History of West Virginia* (n.p., 1913), 547–548.

13. Charles A. Wingerter, "Development of Medical Practice and Public Health," in James Morton Callahan, ed., *Semi-Centennial History of West Virginia* (n.p., 1913), 547.

14. *Transactions of the Medical Society of the State of West Virginia* (Wheeling, 1875), 28.

15. Charles H. Ambler and Festus P. Summers, *West Virginia: The Mountain State*, 2nd ed. (Englewood Cliffs, NJ, 1958), 264–297; Otis K. Rice, *West Virginia: A History* (Lexington, KY, 1985), 154–182; John Alexander Williams, *West Virginia: A Bicentennial History* (New York, 1976), 30–94.

16. *Transactions of the Medical Society of the State of West Virginia* (Wheeling, 1876), 127, 145.

17. Jacob was first elected under the state's original constitution and then reelected under the constitution of 1872. The latter forbade consecutive terms as governor.

18. The Great Railroad Strike of 1877, which wreaked havoc in West Virginia, is mentioned as a key event in virtually every economic history of the United States and every history of American labor published to the present time. The best detailed accounts, however, remain Robert V. Bruce, *1877: Year of Violence* (Indianapolis, 1959); and Philip S. Foner, *The Great Labor Uprising of 1877* (New York, 1977). On federal suppression of the strike, see Jerry M. Cooper, *The Army and Civil Disorder: Federal Military Intervention in Labor Disputes, 1877–1900* (Westport, CT, 1980), 43–98.

19. George Rosen, *A History of Public Health*, expanded ed. (Baltimore, 1993), 224.

20. The AMA's circular letter is reprinted in *Medical and Surgical Reporter*, 36 (1877), 542.

21. On Baird, see Auxiliary to the West Virginia State Medical Association, *Past Presidents of the West Virginia State Medical Association, 1867–1942* (Charleston, 1942), 34–35; and the extensive coverage of his life in the Wheeling papers, following his murder on March 7, 1891.

22. George Baird, "Sanitation in Street Paving," in American Public Health Association, *Public Health Papers and Reports*, vol. 12 (Concord, NH, 1887), 142–144. See also George Baird, "The Destruction of Night Soil and Garbage by Fire," in American Public Health Association, *Public Health Papers and Reports*, vol. 12 (Concord, NH, 1887), 119–122. *Toronto World* (Toronto, ON), October 6, 1886, touts Baird's reputation as a leading health reformer, as does H. J. Sharp, "Report on the Water Supply, Sewage, Etc., of the City of Bellaire,

Ohio," *First Annual Report of the State Board of Health of the State of Ohio for 1886* (Columbus, 1887), 81.

23. *Transactions of the Medical Society of the State of West Virginia* (Wheeling, 1877), 18.

CHAPTER FOUR: *Challenges from Within*

1. *Transactions of the Medical Society of the State of West Virginia* (Wheeling, 1877), 230, 238–239.

2. *Transactions of the Medical Society of the State of West Virginia* (Wheeling, 1879), 449–453.

3. *Transactions of the Medical Society of the State of West Virginia* (Wheeling, 1872), 348–349.

4. Dr. John Frissell Papers, undated scrapbook, West Virginia University Archives and Manuscripts Collection, Box 623, West Virginia University.

5. *Transactions of the Medical Society of the State of West Virginia* (Wheeling, 1879), 450–452.

6. For genealogical information about the Dent family, I am greatly indebted to Rick Dent. Rick kindly corresponded with me on various questions, and his genealogy Web site is the best source of information about the family. See also http://dkwilde.com/Genealogy/Dent/genmain/dent.html, as well as George W. Atkinson and Alvaro F. Gibbens, *Prominent Men of West Virginia* (Wheeling, 1890).

7. *Equal Rights in Religion: Report of the Centennial Congress of Liberals* (Boston, 1876), 26, lists A. M. Dent of Weston, West Virginia, as a contributor.

8. S. T. Wiley, *History of Preston County* (Parsons, WV, [1882] 1993); and scrapbooks of Chuck Plum, a local historian from Newburg, West Virginia, who kindly shared his knowledge of the area.

9. [Anonymous], *Distinguished Successful Americans of Our Day* (Chicago, 1911), 499–501.

10. *Transactions of the Medical Society of the State of West Virginia* (Wheeling, 1881), 679–681.

11. Ibid., 680–681.

12. *Transactions of the Medical Society of the State of West Virginia* (Wheeling, 1882), 721.

13. Sandra Lee Barney, *Authorized to Heal: Gender, Class, and the Transformation of Medicine in Appalachia, 1880–1930* (Chapel Hill, NC, 2000), 30–31.

CHAPTER FIVE: *Securing Legislation*

1. *Wheeling Register* (Wheeling, WV), January 20, 1881; William W. Golden, "The Evolution of Medical Legislation in West Virginia," *West Virginia Medical Journal*, 12, no. 10 (April 1918), 363.

2. James E. Reeves, *The Health and Wealth of the City of Wheeling* (Baltimore, 1871), 139–144. In 1875 he again condemned "foeticide" in a short book en-

titled *The Physical and Moral Causes of Bad Health in American Women* (Wheeling, 1875), 35. But his condemnation in that later book was remarkably low-key, as if he had already decided that the abortion issue lacked the leverage he wanted.

3. James C. Mohr, *Abortion in America: The Origins and Evolution of National Policy, 1800–1900* (New York, 1978).

4. Wilson George Smillie, *Public Health: Its Promise for the Future* (New York, 1955), 309. Some sources, depending on how they assess various state agencies, put the number as high as twenty-three.

5. Geroge Rosen, *A History of Public Health*, expanded ed. (Baltimore, 1993), 223–226.

6. [A Citizen], "The Necessity of a Board of Health," *Wheeling Register* (Wheeling, WV), February 17, 1882.

7. Margaret Humphreys, *Yellow Fever and the South* (New Brunswick, NJ, 1992).

8. *Transactions of the Medical Society of West Virginia, 1882* (Wheeling, 1882), 714–730.

9. *Point Pleasant Register* (Point Pleasant, WV), February 23, 1881.

10. *Transactions of the Medical Society of West Virginia, 1882* (Wheeling, 1882), 726.

11. On Ferguson, see Kenneth R. Bailey, *Alleged Evil Genius: The Life and Times of Judge James H. Ferguson* (Charleston, WV, 2006).

12. See chapter 2, note 13.

13. [Auxiliary to the West Virginia State Medical Association], *Past Presidents of the West Virginia State Medical Association, 1867–1942* (Charleston, WV, 1942), 15–17, 27–28; *Transactions of the Medical Society of West Virginia, 1902* (Wheeling, 1902), 721–723.

14. Alice Jo Hess, *History of Medicine in Harrison County West Virginia* (Parsons, WV, 1978), 556.

15. Sandra Barney, "Bringing Modern Medicine to the Mountains: Scientific Medicine and the Transformation of Health Care in Southern West Virginia, 1880–1910," *West Virginia History*, 55 (1996), 110–126.

16. Quoted in *Border Watchman* (Union, WV), March 24, 1882.

17. Alfred D. Chandler, *The Visible Hand: The Managerial Revolution in American Business* (Cambridge, MA, 1977); Louis Galambos, "The Emerging Organizational Synthesis in Modern American History," *Business History Review*, 44 (Autumn 1970), 270–290; Robert H. Wiebe, *The Search for Order, 1877–1920* (New York, 1967).

18. *Wheeling Register* (Wheeling, WV), January 20, 1881.

19. *Wheeling Register* (Wheeling, WV), January 27, 1881.

20. On William McGrew, see www.mcgrewhouse.org/william-clark-mcgrew.html, accessed May 30, 2012.

21. *Wheeling Register* (Wheeling, WV), February 10, 1881.

22. *State Journal* (Parkersburg, WV), February 17, 1881.

23. *Wheeling Register* (Wheeling, WV), February 22, 1881.

24. On Stollings as railroad capitalist, see *Acts of the Legislature of West Virginia, 1872* (Wheeling, 1872), chapter 15, 25.

25. *Wheeling Register* (Wheeling, WV), February 16, 1881.

26. This was the same William Dawson who later became governor of the state. At that time, he was one of only two Republicans remaining in the West Virginia state senate.

27. *Intelligencer* (Wheeling, WV), February 15, 1881.

28. *Wheeling Register* (Wheeling, WV), February 17, 1881.

29. *Transactions of the Medical Society of the State of West Virginia, 1882* (Wheeling, 1882), 691; George W. Atkinson and Alvaro F. Gibbens, *Prominent Men of West Virginia* (Wheeling, 1890), 485–487.

30. *Intelligencer*, (Wheeling, WV), February 23, 1881.

31. Alexander Wilder, ed., *Transactions of the National Eclectic Medical Association*, vol. 17 (1889–1890), 59–60; Thomas L. Bradford, "Homoeopathy in West Virginia," in William H. King, *History of Homoeopathy and Its Institutions in America; Their Founders, Benefactors, Faculties, Officers, Hospitals, Alumni, etc., with a Record of Achievement of its Representatives in the World of Medicine* (New York, 1905), vol. 1, 402.

32. William G. Rothstein, *American Physicians in the Nineteenth Century: From Sects to Science* (Baltimore, 1972), 152–197, 217–246.

33. *Wheeling Register* (Wheeling, WV), March 9, 1881.

34. *Transactions of the Medical Society of the State of West Virginia, 1882* (Wheeling, 1882), 720.

35. *Ritchie Gazette* (Harrisburg, WV), March 17, 1881; *Wheeling Register* (Wheeling, WV), March 9, 1881; *Preston County Journal* (Kingwood, WV), March 17, 1881.

36. Charles H. Ambler and Festus P. Summers, *West Virginia: the Mountain State*, 2nd ed. (Englewood Cliffs, NJ, 1958), 284.

37. *Wheeling Register* (Wheeling, WV), March 10, 1881.

38. *Transactions of the Medical Society of the State of West Virginia, 1882* (Wheeling, 1882), 667.

39. William W. Golden, "The Evolution of Medical Legislation in West Virginia," *West Virginia Medical Journal*, 12, no. 10 (April 1918), 363.

CHAPTER SIX: *Exercising Power*

1. Jackson had already replaced Dr. T. B. Camden as superintendent of the state insane asylum, thereby signaling his determination to appoint medical officers who shared his views just as promptly as he could. See Camden Papers, Box 623, West Virginia Archives and Manuscript Collections, West Virginia University.

2. *First, Second and Third Annual Reports of the Secretary of the State Board of Health of West Virginia for the Years Ending December 31st, 1881, 1882, 1883* (Wheeling, 1884), 25.

3. *Transactions of the Medical Society of the State of West Virginia* (Wheeling, 1882), 725.

4. Ibid., 720, 721.

5. *Point Pleasant Register* (Point Pleasant, WV), February 9, 1881.

6. Minutes of the board's initial session appear in *First, Second and Third Annual Reports of the Secretary of the State Board of Health of West Virginia for the Years Ending December 31st, 1881, 1882, 1883* (Wheeling, 1884), 3–30.

7. Ibid., 32.

8. Ibid., 27–30.

9. Ibid., 29–30; *Kentucky Medical Journal*, 11 (July 1913), 578; *Journal of the American Medical Association*, 64, no. 21 (May, 1915), 1796; and Hopkins's obituary at www.rootsweb.ancestry.com/~kypendle/Pages/obitvarious.htm, accessed November 3, 2010.

10. *First, Second and Third Annual Reports of the Secretary of the State Board of Health of West Virginia for the Years Ending December 31st, 1881, 1882, 1883* (Wheeling, 1884), 32–33.

11. *State Journal* (Parkersburg, WV), March 3, 1881.

12. *Wheeling Register* (Wheeling, WV), January 12, 1882.

13. Ibid.; *Proceedings of the American Pharmaceutical Association*, vol. 43 (Baltimore, 1895), 41–42.

14. *Transactions of the Medical Society of the State of West Virginia* (Wheeling, 1882), 720.

15. *Wheeling Register* (Wheeling, WV), January 19, 1882.

16. *Transactions of the Medical Society of the State of West Virginia* (Wheeling, 1882), 720–721.

17. House Bill no. 405, in *Journal of the House of Delegates of the State of West Virginia for the Adjourned Session, 1882* (Wheeling, 1882).

18. *Wheeling Register* (Wheeling, WV), February 14, 1882; *Journal of the House of Delegates of the State of West Virginia for the Adjourned Session, 1882* (Wheeling, 1882), 223–224.

19. *Wheeling Daily Intelligencer* (Wheeling, WV), February 14, 1882.

20. *Statutes of West Virginia, 1882*, chapter 93, section 9.

21. *Wheeling Daily Intelligencer* (Wheeling, WV), February 15, 1882; *Border Watchman* (Union, WV), February 24, 1882.

22. *Journal of the House of Delegates of the State of West Virginia for the Adjourned Session, 1882* (Wheeling, 1882), 224.

23. *Wheeling Register* (Wheeling, WV), February 14, 1882.

24. *Independent* (Martinsburg, WV), February 18, 1882.

25. *Wheeling Daily Intelligencer* (Wheeling, WV), March 4 and 6, 1882.

26. *Journal of the House of Delegates of the State of West Virginia for the Adjourned Session, 1882* (Wheeling, 1882), 400–402, 457–458, 470–471.

27. *Transactions of the Medical Society of the State of West Virginia* (Wheeling, 1882), 720.

28. "Necrology: James Edmund Reeves," *Journal of the American Medical Association*, 26 (January 18, 1896), 142.

CHAPTER SEVEN: *The Dents Confront the Board*

1. *Wheeling Intelligencer* (Wheeling, WV), March 16, 1882.

2. The tangled tale of A. M. Dent and the Columbus Medical College has been assembled principally from D. N. Kinsman, "Medical Education in Ohio," *Medical News*, 41 (July 1882), 109–110; an editorial column in *Medical News*, 42 (April 1883), 420; and "Columbus Medical College Imbroglio," *Gaillard's Medical Journal*, 34 (1882), 297–298. Other sources will be cited separately.

3. *Weston Democrat* (Weston, WV), reprinted in the *New Dominion* (Morgantown, WV), March 11, 1882.

4. *Proceedings of the State Board of Health* (July 1881), 36–37.

5. *Proceedings of the State Board of Health* (January 1882), 40.

6. United States Census, 1880, Preston County, West Virginia.

7. *Fourth Annual Report of the Secretary of the State Board of Health of West Virginia for the Year Ending December 31st, 1884* (Wheeling, 1884), 21–26.

8. John S. Haller Jr., *A Profile in Alternative Medicine: The Eclectic Medical College of Cincinnati, 1845–1942* (Kent, OH, 1999), is the definitive scholarly study of Eclectic medical education in Cincinnati; for Reeves's accusations, see "Will the Truth Overtake a Lie?," *Eclectic Medical Journal*, 45 (1885), 96.

9. On Lanham and his affiliations, see *Acts of the Legislature of West Virginia, 1889* (Charleston, 1889), 483; *Journal of the American Medical Association*, 37, no. 1 (July 6, 1901), 36; *Twenty-Sixth Biennial Report of the Attorney General of the State of West Virginia* (Charleston, 1916), 285–286; and *Annual Report of the State Health Department of West Virginia* (Charleston, 1920), 14.

10. *Statutes of the State of West Virginia, 1882*, chapter 93, section 9.

11. "Society Proceedings," in *Medical News* (NY), 41 (October 28, 1882).

12. *Fourth Annual Report of the Secretary of the State Board of Health of West Virginia for the Year Ending December 31st, 1884* (Wheeling, 1884), 26.

13. The University of West Virginia Alumni Association created the Marmaduke Dent Society in his honor in 1993. The society recognizes donors in the $5,000 to $50,000 range. See http://alumni.wvu.edu/awards/marmaduke, accessed February 21, 2011.

14. John Philip Reid, *An American Judge: Marmaduke Dent of West Virginia* (New York, 1968).

15. *Preston County Journal* (Kingwood, WV), December 7, 1882.

16. *Court Ledgers*, Preston County Circuit Court, Kingwood, WV, August 18, November 6, and December 14, 1882.

17. *Preston County Journal* (Kingwood, WV), April 19, 1883.

CHAPTER EIGHT: *The West Virginia State Supreme Court*

1. *West Virginia v. Dent* [25 W.Va., 1:1884], in the boxed archives of the Supreme Court of Appeals of West Virginia, West Virginia School of Law, Morgantown, WV. I am greatly indebted to the staff who helped me find these records and gave me access to them.

2. The petition was reprinted in *Fourth Annual Report of the Secretary of the State Board of Health of West Virginia, for the Year Ending December 31st, 1884* (Wheeling, 1884), 21–26.

3. Ibid.

4. A state constitutional amendment in 1904 subsequently increased the number of justices to the present five.

5. "Judges of the Supreme Court of Appeals during the Time of These Reports," *Reports of Cases Argued and Determined in the Supreme Court of Appeals of West Virginia at the Fall-Special Term of 1884, and the January Spring-Special Term of 1885*, vol. 25 (1884–1885), iii. Johnson presided as "president" of the court for these special sessions.

6. These quotes and the ones that follow from Marmaduke's brief can be found in a privately printed pamphlet entitled *In the Supreme Court of Appeals of West Virginia, Brief and Argument for Appellant*, in the boxed archives of the Supreme Court of Appeals of West Virginia, West Virginia School of Law, Morgantown, WV.

7. For examples of such legislation, see James C. Mohr, *Doctors and the Law: Medical Jurisprudence in Nineteenth-Century America* (New York, 1993), 76–93; James C. Mohr, "American Physicians and the Criminal Law," International Congress for the History of Medicine, Galveston, TX (2000); James C. Mohr, "The Regulation of Medical Ethics in American History," plenary address for the international conference on bio-ethics, University of North Dakota School of Medicine and Health Sciences (2005); and Stephen R. Latham and James C. Mohr, "The Legal and Quasilegal Regulation of Practitioners and Practice in the United States," in Robert B. Baker and Laurence B. McCullough, eds., *The Cambridge World History of Medical Ethics* (New York, 2009), 540–551.

8. Garfield's personal physician, Doctor Willard Bliss (his given name was Doctor), was a distinguished Regular MD who had earlier been one of the bedside attendants when President Abraham Lincoln died. Bliss's handling of Garfield's case was roundly condemned by the international medical press at the time and by medical historians ever since. For the most recent account, see Candice Millard, *Destiny of the Republic: A Tale of Madness, Medicine, and Murder of a President* (New York, 2011).

9. *Cummings v. Missouri*, 71 U.S. 277 (1867).

10. *Ex parte Garland*, 71 U.S. 333 (1866).

11. On C. C. Watts, see George Wesley Atkinson and Alvaro Franklin Gibbens, *Prominent Men of West Virginia* (Wheeling, WV, 1890), 812–813; and

http://politicalgraveyard.com/bio/watts.html, accessed March 29, 2011. Watts would later become United States Attorney for West Virginia during the Cleveland administration and ran unsuccessfully as the Democratic nominee for governor in 1896.

12. The court's own box of records from this case does not contain Watts's brief.

13. Green's complete opinion can be found in *Reports of Cases Argued and Determined in the Supreme Court of Appeals of West Virginia*, vol. 25 (Charleston, WV, 1902), 1–23.

14. *Wheeling Register* (Wheeling, WV), November 2, 1884; *Intelligencer* (Wheeling, WV), November 2, 1884; *Preston County Journal* (Kingwood, WV), November 6, 1884.

15. *Medical News* (NY), 45 (December 6, 1884), 630–631.

16. James William Moore, *Moore's Federal Practice*, ed. Daniel R. Coquillette et al. (New York, 1997), vol. 12, sections 400.05-400.07. At that time, defense attorneys could obtain a writ of error from the lower court judge, as Marmaduke did, or from a federal judge.

CHAPTER NINE: *Conflict and Enforcement*

1. *First, Second and Third Annual Reports of the Secretary of the State Board of Health of West Virginia for the Years Ending December 31st, 1881, 1882, 1883* (Wheeling, 1883), 32.

2. Ibid., 45.

3. *Fourth Annual Report of the Secretary of the State Board of Health of West Virginia for the Year Ending December 31st, 1884* (Wheeling, 1884), 21.

4. *Wheeling Register* (Wheeling, WV), November 2, 1884.

5. *First, Second and Third Annual Reports of the Secretary of the State Board of Health of West Virginia for the Years Ending December 31st, 1881, 1882, 1883* (Wheeling, 1883), 36–37.

6. *Transactions of the Medical Society of West Virginia, 1883*, 17, 20–21, 26–33. On Allen personally, see [Auxiliary to the West Virginia State Medical Association], *Past Presidents of the West Virginia State Medical Association, 1867–1942* (Charleston, WV, 1942), 30–32.

7. *Philadelphia Medical Times*, 13 (July 14, 1883), 727.

8. *Transactions of the Medical Society of West Virginia, 1896*, 1324.

9. *Intelligencer* (Wheeling, WV), July 18, 1887.

10. *Transactions of the Medical Society of West Virginia, 1885*, 167–185.

11. The quote is from Otis K. Rice, *West Virginia: A History* (Lexington, KY, 1985), 171–172. See also Charles H. Ambler and Festus P. Summers, *West Virginia: the Mountain State*, 2nd ed. (Englewood Cliffs, NJ, 1958), 284–285; John Alexander Williams, *West Virginia: A Bicentennial History* (New York, 1976), 122–123.

12. *Wheeling Register* (Wheeling, WV), February 14, 1882; *Journal of the House of Delegates of the State of West Virginia, 1882* (Wheeling, 1882), 223–224.

13. *Intelligencer* (Wheeling, WV), July 18, 1887.

14. On Garrison's background, see [Linda Fluharty], "George I. Garrison," in *History of the Upper Ohio Valley*, vol. 1 (n.p., 1890), 295–296. For the incessant public feuding between Garrison, on the one hand, and Baird, the city council, and the medical establishment, on the other, see Elan Ebeling, "The Tumultuous Nature of American Public Health at the Grass Roots Level during a Transitional Decade: Wheeling, West Virginia, 1880–1890" (senior thesis, Clark Honors College, University of Oregon, 2011).

15. *Transactions of the Medical Society of West Virginia, 1887*, 359.

16. *Journal of the American Medical Association*, 9 (July 1887), 151; "Jefferson College Graduates a Student with One Course of Lectures," *Eclectic Medical Journal*, 67 (1887), 478–479; *Intelligencer* (Wheeling, WV), July 18, 1887.

17. Jefferson Medical College, "Requirements for Graduation," *Annual Announcement, 1885–6*, 17. I am indebted to Michael Angelo for providing a copy of this document and the records concerning Garrison's attendance and degree at JMC.

18. *Intelligencer* (Wheeling, WV), August 18, 1887.

19. The correspondence was published in *Biennial Report of the State Board of Health, of the State of West Virginia, for the Years 1885 and 1886* (Wheeling, 1886), 19–21.

20. *Biennial Report of the State Board of Health, of the State of West Virginia, for the Years 1887 and 1888* (Wheeling, 1888), 15, 21, and passim.

21. Frances Priscilla De Lancy, *The Licensing of Professions in West Virginia* (Chicago, 1938), 26.

22. Supreme Court of Appeals of West Virginia, *State vs. Ragland*, reprinted in *Biennial Report of the State Board of Health, of the State of West Virginia, for the Years 1887 and 1888* (Wheeling, 1888), 65–69. See also Sandra Lee Barney, *Authorized to Heal: Gender, Class, and the Transformation of Medicine in Appalachia, 1880–1930* (Chapel Hill, NC, 2000), 27.

CHAPTER TEN: *The United States Supreme Court*

1. James William Moore, *Moore's Federal Practice*, ed. Daniel R. Coquillette et al. (New York, 1997), vol. 12, sections 400.05–400.07.

2. On Miller's medical background, see Michael A. Ross, *Justice of Shattered Dreams: Samuel Freeman Miller and the Supreme Court during the Civil War Era* (Baton Rouge, 2003), 1–18; Michael A. Ross, "Justice Miller's Reconstruction: *The Slaughter-House Cases*, Health Codes, and Civil Rights in New Orleans, 1861–1873," *Journal of Southern History*, 64, no. 4 (November 1998), 649–676; and Wendy E. Parmet, "From Slaughter-House to Lochner: The Rise and Fall of the Constitutionalization of Public Health," *American Journal of Legal History*, 40 (October 1996), 476–505.

3. On Blatchford's uncle, see James C. Mohr, *Abortion in America: The Origins and Evolution of National Policy, 1800–1900* (New York, 1978), 154–155.

4. On Field's cholera incident, see his own memoir, *California Alcalde* (Oakland, CA, 1950), 15.

5. Marmaduke's arguments in this and the following paragraphs are taken from "Argument in Behalf of Plaintiff in Error," *Transcript of Record, Supreme Court of the United States, October Term, 1888, Frank M. Dent, plaintiff in error vs. the State of West Virginia* (filed October 13, 1888), 1–6.

6. Caldwell's arguments in this and the following paragraphs are taken from "Brief for the Defendant in Error," *Transcript of Record, Supreme Court of the United States, October Term, 1888, Frank M. Dent, plaintiff in error vs. the State of West Virginia* (filed October 13, 1888), 1–19.

7. On Field's much-analyzed career, see Philip J. Bergan, Owen M. Fiss, and Charles W. McCurdy, *The Fields and the Law* (New York, 1986); Charles W. McCurdy, *Economic Growth and Judicial Conservatism in the Age of Enterprise: Studies in the Jurisprudence of Stephen J. Field, 1850–1900* (Ann Arbor, MI, 1976); Charles W. McCurdy, "Justice Field and the Jurisprudence of Government-Business Relations: Some Parameters of Laissez-Faire Constitutionalism, 1863–1897," *Journal of American History*, 61 (March 1975), 970–1005; Carl Brent Swisher, *Stephen J. Field: Craftsman of the Law* (Washington, DC, 1930); and Paul Kens, *Justice Stephen Field: Shaping Liberty from the Gold Rush to the Gilded Age* (Lawrence, KS, 1997).

8. *Slaughter-House Cases*, 83 U.S. 36. See also Ronald M. Labbé and Jonathan Lurie, *The Slaughterhouse Cases: Regulation, Reconstruction, and the Fourteenth Amendment* (Lawrence, KS, 2003).

9. The quote is from Gary D. Rowe's masterful summary of contemporary interpretations of the Supreme Court during this period: "*Lochner* Revisionism Revisited," *Law and Social Inquiry*, 24 (1999), 221–252 (quote p. 226, parentheses in original). The characterization of the court offered here rests primarily upon the work of Alan Jones, "Thomas M. Cooley and 'Laissez-Faire Constitutionalism': A Reconsideration," *Journal of American History*, 53 (1967), 751–771; Charles W. McCurdy, "Justice Field and the Jurisprudence of Government-Business Relations," *Journal of American History*, 61 (1975), 970–1005; Michael Les Benedict, "Laissez-Faire and Liberty: A Re-Evaluation of the Meaning and Origins of Laissez-Faire Constitutionalism," *Law and History Review*, 3 (1985), 293–331; Morton J. Horwitz, *The Transformation of American Law, 1870–1960: The Crisis of Legal Orthodoxy* (New York, 1992); Howard Gillman, *The Constitution Besieged: The Rise and Demise of Locher Era Police Powers Jurisprudence* (Durham, NC, 1993); Owen M. Fiss, *History of the Supreme Court of the United States: Troubled Beginnings of the Modern State, 1888–1910* [vol. 8 of the Holmes Devise history of the Supreme Court] (New York, 1993); Barbara Young Welke, *Recasting American Liberty: Gender, Race, Law, and the Railroad Revolution, 1865–1920* (New York, 2001); James W. Ely, *The Chief Justiceship of Melville W. Fuller* (Columbia, SC, 1995); Marcus D. Dubber and Mariana Valverde, eds., *Police and the Liberal State* (Stanford, CA, 2008); and Brian Balogh, *A Government Out of Sight: The Mystery of National Authority in Nineteenth-Century America* (New York, 2009). Even David E. Bernstein, who argues that the court's driving principle was the

defense of individual liberties, seems to concede that the justices were more attuned to issues of class legislation and abuses of power by special interests through the 1880s than they later became by the first decade of the twentieth century. See David E. Bernstein, *Rehabilitating Lochner: Defending Individual Rights against Progressive Reform* (Chicago, 2011), 14–15, and passim; and the critique of Bernstein's position in Barry Cushman, "Some Varieties and Vicissitudes of Lochnerism," *Boston University Law Review*, 85 (2005), 881–1000. William J. Novak argues persuasively in "Law and the Social Control of American Capitalism," *Emory Law Journal*, 60, no. 2 (2010), 377–405, that the court was more open to restrictive regulations during the period 1877–1932 than most scholars seem to realize, but he deals principally with regulations affecting what he characterizes as "public utility" issues, and his most compelling examples come after the *Dent* case.

10. *Butchers' Union Slaughter-house and Live-stock Landing Co. v. Crescent City Live-stock Landing and Slaughter-house Co.*, 111 U.S. 757 (1884).

11. *Butchers' Union Slaughter-house and Live-stock Landing Co. v. Crescent City Live-stock Landing and Slaughter-house Co.*, 111 U.S. 766 (1884).

12. On the approval of previous regulations as reasonable, see David E. Bernstein, *Rehabilitating Lochner: Defending Individual Rights against Progressive Reform* (Chicago, 2011), 15, and note 57, 137.

13. *Yick Wo v. Hopkins, Sheriff of San Francisco*, 118 U.S. 353. At least one early authority on due process believed that the court's choice of the equal protection in that case was essentially arbitrary, since the justices had not articulated a sharp distinction between equal protection and due process; they could just as easily have invoked due process. See Rodney L. Mott, *Due Process of Law: A Historical and Analytical Treatise of the Principles and Methods Followed by Courts in the Application of the Concept of the "Law of the Land"* (New York, [1926] 1973), 284–289.

14. These and all subsequent quotes from Field's decision are from *Dent v. West Virginia*, 129 U.S. 114 (1889). For an excellent discussion of the right to pursue ordinary callings and common trades, see James W. Ely Jr., "'To Pursue Any Lawful Trade or Occupation': The Evolution of Unenumerated Economic Rights in the Nineteenth Century," *Journal of Constitutional Law*, 8, no. 25 (2006), 917–955.

15. James W. Ely Jr., "'To Pursue Any Lawful Trade or Occupation': The Evolution of Unenumerated Economic Rights in the Nineteenth Century," *Journal of Constitutional Law*, 8, no. 25 (2006), 947; Lawrence M. Friedman, "Freedom of Contract and Occupational Licensing, 1890–1910: A Legal and Social Study," *California Law Review*, 53 (May 1965), 487–534.

CHAPTER ELEVEN: *American Medical Practice after* Dent

1. Stephen R. Latham and James C. Mohr, "The Legal and Quasilegal Regulation of Practitioners and Practice in the United States," in Robert B. Baker and Laurence B. McCullough, eds., *The Cambridge World History of Medical*

Ethics (New York, 2009), 543–544; and James C. Mohr, "The Regulation of Medical Ethics in American History," plenary address for national conference on bio-ethics, University of North Dakota School of Medicine and Health Sciences (2005).

2. See, among many others, the long-standard narrative by Fielding H. Garrison, *An Introduction to the History of Medicine* (Philadelphia, 1913, and subsequent editions); Richard Harrison Shryock, *Medical Licensing in America, 1650–1965* (Baltimore, 1967), especially 109; and the theoretical arguments embodied in the influential work of Joseph Ben-David, "Roles and Innovations in Medicine," *American Journal of Sociology*, 65 (May 1960), 557–568. Even so perceptive a scholar as George Rosen accepted this general line of analysis: see George Rosen, *The Structure of American Medical Practice, 1875–1941*, ed. Charles E. Rosenberg (Philadelphia, 1983). On related issues, see the essays in Frank Huisman and John Harley Warner, eds., *Locating Medical History: The Stories and Their Meanings* (Baltimore, 2004).

3. Gerald L. Geison, "Divided We Stand: Physiologists and Clinicians in the American Context," in Morris J. Vogel and Charles E. Rosenberg, eds., *The Therapeutic Revolution: Essays in the Social History of American Medicine* (Philadelphia, 1979), 67–90; John Harley Warner, *The Therapeutic Perspective: Medical Practice, Knowledge, and Identity in America, 1820–1885* (Cambridge, MA, 1986); Ronald L. Numbers and John Harley Warner, "The Maturation of American Medical Science," in Nathan Reingold and Marc Rothenberg, eds., *Scientific Colonialism: A Cross-Cultural Comparison* (Washington, DC, 1987), 191–214.

4. William G. Rothstein, "When Did a Random Patient Benefit from a Random Physician? Introduction and Historical Background"; Rosemary Stevens, "Technology and Institutions in the Twentieth Century"; Dale C. Smith, "Surgery: It's Not a Random Therapy"; Steven J. Peitzman, "When Did Medicine Become Beneficial? The Perspective from Internal Medicine"; and John Parascandola, "Drug Therapy and the Random Patient"; all in William G. Rothstein, ed., *CADUCEUS: A Humanities Journal for Medicine and the Health Sciences*, 12, no. 3 (1996). There are a few exceptions to this generalization, including perhaps most importantly the development of a diphtheria antitoxin in 1894, along with some early twentieth-century hormone treatments. But the generalization remains largely true for most medical circumstances and most common diseases.

5. *Transactions of the Medical Society of the State of West Virginia, 1882* (Wheeling, 1882), 718.

6. *Medical News*, 54 (January 26, 1889), 104.

7. *North Carolina Medical Journal*, 25 (1890), 215–217, 224–226.

8. Reginald H. Fitz, *The Legislative Control of Medical Practice* (Boston, 1894); Howard A. Kelly and Walter L. Burrage, *American Medical Biographies* (Baltimore, 1920), 389–390.

9. "The Legislative Control of Medical Practice," *Atlanta Medical and Surgical Journal*, 11 (August 1894), 351–354.

10. For an early example of this trend that caught national attention, see *Indiana Medical Journal*, 15 (1896–1897), 327–328.

11. Alexander Wilder, ed., *Transactions of the National Eclectic Medical Association*, 17 (1889–1890), 59–60, and passim.

12. William G. Rothstein, *American Physicians in the Nineteenth Century: From Sects to Science* (Baltimore, 1972), table 16.1, 308.

13. Alan Jones, "Thomas M. Cooley and 'Laissez-Faire Constitutionalism': A Reconsideration," *Journal of American History*, 53 (1967), 751–771; Charles W. McCurdy, "Justice Field and the Jurisprudence of Government-Business Relations," *Journal of American History*, 61 (1975), 970–1005; Michael Les Benedict, "Laissez-Faire and Liberty: A Re-Evaluation of the Meaning and Origins of Laissez-Faire Constitutionalism," *Law and History Review*, 3 (1985), 293–331; Morton J. Horwitz, *The Transformation of American Law, 1870–1960: The Crisis of Legal Orthodoxy* (New York, 1992); Howard Gillman, *The Constitution Besieged: The Rise and Demise of Locher Era Police Powers Jurisprudence* (Durham, NC, 1993); Owen M. Fiss, *History of the Supreme Court of the United States: Troubled Beginnings of the Modern State, 1888–1910* [vol. 8 of the Holmes Devise history of the Supreme Court] (New York, 1993); Gary D. Rowe, "*Lochner* Revisionism Revisited," *Law and Social Inquiry*, 24 (1999), 221–252; Barbara Young Welke, *Recasting American Liberty: Gender, Race, Law, and the Railroad Revolution, 1865–1920* (New York, 2001); James W. Ely, *The Chief Justiceship of Melville W. Fuller* (Columbia, SC, 1995); Marcus D. Dubber and Mariana Valverde, eds., *Police and the Liberal State* (Stanford, CA, 2008); and Brian Balogh, *A Government Out of Sight: The Mystery of National Authority in Nineteenth-Century America* (New York, 2009). Even David E. Bernstein, who argues that the court's driving principle was the defense of individual liberties, seems to concede that the justices were more attuned to issues of class legislation and abuses of power by special interests through the 1880s than they later became by the first decade of the twentieth century. See David E. Bernstein, *Rehabilitating Lochner: Defending Individual Rights against Progressive Reform* (Chicago, 2011), 14–15, and passim; and the critique of Bernstein's position in Barry Cushman, "Some Varieties and Vicissitudes of Lochnerism," *Boston University Law Review*, 85 (2005), 881–1000.

14. S. S. Merrill, "Note," *Central Law Journal*, 28, no. 11 (March 15, 1889), 266.

15. George Hoadly, "The Constitutional Guarantees of the Right of Property as Affected by Recent Decisions," *Journal of Social Sciences*, 26 (1890), 5–11. On the ASSA, see Thomas L. Haskell, *The Emergence of Professional Social Science: The American Social Science Association and the Nineteenth-Century Crisis of Authority* (Urbana, IL, 1977).

16. Thomas M. Cooley, *A Treatise on the Constitutional Limitations Which Rest Upon the Legislative Power of the States of the American Union*, 6th ed. (Boston, 1890), 16, 437.

17. Thomas M. Cooley, *A Treatise on the Constitutional Limitations Which Rest Upon the Legislative Power of the States of the American Union*, 7th ed. (Boston, 1903), 19, 376, 509.

18. J. W. Meyers, "The Privileges and Immunities of Citizens in the Several States," *Michigan Law Review* (1902–1903), 294.

19. Ernst Freund, *The Police Power, Public Policy and Constitutional Rights* (Chicago, 1904), 571–573.

20. Isaac Franklin Russell, "Due Process of Law," *Yale Law Review*, 14 (1905), 331.

21. Horace W. Fuller et al., *The Green Bag: A Useless but Entertaining Magazine for Lawyers* (1908), 597.

22. *Dent v. West Virginia*, 129 U.S. 114 (1889).

23. For the court's previous use of the phrase "ordinary avocations," see *Ex parte Garland*, 71 U.S. 333 (1866); *Slaughter-House Cases*, 83 U.S. 36 (1873); *Bradwell v. Illinois*, 83 U.S. 130 (1873). For an excellent discussion of this issue in the years after the *Dent* decision, see James W. Ely Jr., "'To Pursue Any Lawful Trade or Occupation': The Evolution of Unenumerated Economic Rights in the Nineteenth Century," *Journal of Constitutional Law*, 8, no. 5 (2006), 917–955.

24. For the ambiguities involved in the question of whether the doctor/patient relationship amounted to a de facto contract, see Kenneth A. DeVille, *Medical Malpractice in Nineteenth-Century America: Origins and Legacy* (New York, 1990), 156–186.

25. This principle had already been established for the profession of law. In *Bradwell v. Illinois*, 83 U.S. 130 (1873), the court ruled against a woman who invoked the "privileges and immunities" clause of the Fourteenth Amendment in her effort to break a ban on females practicing law in that jurisdiction.

26. C. F. Ulrich, "A Few Hints Relative to Elevating the Standard of the Medical Profession," *Transactions of the Medical Society of the State of West Virginia* (Wheeling, 1881), 691.

27. Many legal scholars point out that American courts, including the Supreme Court, subsequently approved many other regulatory statutes, thereby enabling the modern regulatory state. But most of the decisions they cite came after *Dent* and involved general principles of economic regulation as distinguished from occupational exclusion. For variations on this argument, see Felice Batlan, "A Reevaluation of the New York Court of Appeals: The Home, the Market, and Labor, 1885–1905," *Law and Social Inquiry*, 27, no. 3 (Summer 2002), 489–528; Wendy E. Parmet, "From Slaughter-House to Lochner: The Rise and Fall of the Constitutionalization of Public Health," *American Journal of Legal History*, 40, no. 4 (October 1996), 476–505; and the works cited in chapter 10, note 9.

28. Rodney L. Mott, *Due Process of Law: A Historical and Analytical Treatise of the Principles and Methods Followed by the Courts in the Application of the Concept of the "Law of the Land"* (New York, 1973 [reprint of 1926 ed.]), 246–287, 340–342. In this context, see also Barry Cushman, "Some Varieties and Vicissitudes of Lochnerism," *Boston University Law Review*, 85 (2005), 885–886; and David E. Bernstein, *Rehabilitating Lochner: Defending Individual Rights against Progressive Reform* (Chicago, 2011), 15.

29. Lawrence M. Friedman, "Freedom of Contract and Occupational Licensing, 1890–1910: A Legal and Social Study," *California Law Review*, 53 (May 1965), 487–534, argued that medical licenses were the first in a subsequent line of occupational regulations that he called "friendly" licenses, by which he meant successful efforts on the part of those theoretically being regulated to use the state to rationalize and control their own marketplaces. In that context, the early licensing of doctors would not seem radical; if barbers could be licensed on the pretext of protecting public health and safety, surely physicians should be. I agree with Friedman that medical licenses were quintessential examples of "friendly" licenses in his terms. But I do not believe they were essentially similar to the many other friendly licenses that followed through the twentieth century. Unlike medical licenses, all of the later friendly licenses—even those justified on health grounds—were based upon relatively measurable and objective criteria, and they could be lost for cause. Licensed morticians, for example, were subject to outside inspection, and if they failed to maintain prescribed standards, whether anyone was harmed or not in a legal sense, the state could revoke their licenses. Moreover, the holders of subsequent friendly licenses also remained responsible for the outcomes of what they did and hence were not treated in the eyes of the law as true professionals. Though licensed, they remained "ordinary occupations."

30. On nineteenth-century efforts to set fees, see Donald E. Konold, *A History of American Medical Ethics, 1847–1912* (Madison, WI, 1962); and Joseph F. Kett, *The Formation of the American Medical Profession: the Role of Institutions, 1780–1860* (New Haven, 1968). On the comparative lack of regulation in the United States, see Richard Harrison Shryock, *Medical Licensing in America, 1650–1965* (Baltimore, 1967).

31. J. C. Irons, "Should Not Legislative Enactment Authorize the Revoking of State Medical License?," *Transactions of the Medical Society of West Virginia* (1896), 1375–1377. On Irons personally, see [American Historical Association], *The History of West Virginia, Old and New*, vol. 3 (New York, 1923), 558. Another politically active physician, Irons subsequently became the first mayor of Elkins.

32. AMA policy continues to insist that "the medical profession alone [i.e., neither state boards nor state legislatures] has the responsibility for setting standards and determining curricula in continuing medical education." Carolyne Krupa, "Hot-Button Issues Drive State CME Mandates," *American Medical News*, posted February 13, 2012.

33. It might be argued that physicians were not exceptional in having preparation-only licenses; lawyers, for example, passed the bar once and were thereafter free to practice in that jurisdiction regardless of whether they ever read another treatise. But lawyers competed with each other directly, in largely public forums, and the results of their competitions were known to interested parties. Consequently, they had a powerful incentive to remain as sharp as they could, since ineffective or incompetent lawyers would lose to well-prepared lawyers most of the time and would soon find themselves without clients. Competition also tended to restrain lawyers' fees. That was not the case with physicians. Physicians did not compete with one another the way lawyers did (in fact, they generally agreed not to compete), and the public had no way to judge their relative effectiveness or ongoing competence, since every physician's patient outcomes remained private—indeed, private by law.

34. On revocations, see, among many others, Morrison Wickersham, "Physicians Disciplined by a State Medical Board," *Journal of the American Medical Association*, 307, no. 3 (January 18, 2012), 223–230; and Roberto Cardarelli and John C. Licciadone, "Factors Associated with High-Severity Disciplinary Action by a State Medical Board: A Texas Study of Medical License Revocation," *Journal of the American Osteopathic Association*, 106, no. 3 (March 1, 2006), 153–156. Even Richard Harrison Shryock, otherwise a champion of medical licensing as a progressive reform, was dubious about the profession's self-policing through the time he was writing in the 1960s. See Richard Harrison Shryock, *Medical Licensing in America, 1650–1965* (Baltimore, 1967), 114. Carl F. Ameringer, *State Medical Boards and the Politics of Public Protection* (Baltimore, 1999); and David A. Johnson and Humayun J. Chaudry, *Medical Licensing and Discipline in America: A History of the Federation of State Medical Boards* (Lanham, MD, 2012) discuss efforts to reform the nation's medical oversight.

35. Linda T. Kohn, Janet M. Corrigan, and Molla S. Donaldson, eds., Committee on Quality of Health Care in America, Institute of Medicine, *To Err Is Human: Building a Safer Health System* (Washington, DC, 2000).

36. On the history and evolution of medical malpractice litigation, see James C. Mohr, "American Medical Malpractice Litigation in Historical Perspective," *Journal of the American Medical Association*, 283 (April 5, 2000), 1731–1737; and Kenneth Allen De Ville, *Medical Malpractice in Nineteenth-Century America: Origins and Legacy* (New York, 1990). The literature on the modern malpractice system is both enormous and contentious. Readers may want to start with the following classic studies: [President and Fellows of Harvard College], *Patients, Doctors, and Lawyers: Medical Injury, Malpractice Litigation, and Patient Compensation in New York* (Cambridge, MA, 1990); Paul C. Weiler, *Medical Malpractice on Trial* (Cambridge, MA, 1991); Paul C. Weiler et al., *A Measure of Malpractice: Medical Injury, Malpractice Litigation, and Patient Compensation* (Cambridge, MA, 1993); United States Congress, Office of Technology Assessment, *Defensive Medicine and Medical Malpractice*

(Washington, DC, 1994); Troyen A. Brennan et al., "Relation between Negligent Adverse Events and the Outcomes of Medical-Malpractice Litigation," *New England Journal of Medicine*, 335 (1996), 1963–1967; and Neil Vidmar, *Medical Malpractice and the American Jury: Confronting the Myths about Jury Incompetence, Deep Pockets, and Outrageous Damage Awards* (Ann Arbor, MI, 1995).

37. Hugh Emmett Culbertson, *Medical Men and the Law: A Modern Treatise on the Legal Rights Duties and Liabilities of Physicians and Surgeons* (Philadelphia and New York, 1913), 34.

38. See forthcoming dissertation by Clinton Sandvick, University of Oregon.

39. Frances Priscilla De Lancy, *The Licensing of Professions in West Virginia* (Chicago, 1938), 25.

40. The consolidation of the American medical profession during the first third of the twentieth century has been well documented by several scholars. A fine summary of the process may be found in Ronald L. Numbers, "The Fall and Rise of the American Medical Profession," in Nathan O. Hatch, ed., *The Professions in American History* (Notre Dame, IN 1988), 51–72. William G. Rothstein, *American Medical Schools and the Practice of Medicine* (New York, 1987), and Kenneth M. Ludmerer, *Learning to Heal: The Development of American Medical Education* (New York, 1985), document the consolidation of medical education that underlay the larger process.

41. Frances Priscilla De Lancy, *The Licensing of Professions in West Virginia* (Chicago, 1938), 24. That law was further strengthened in 1929.

42. On the lack of enforcement through the first half of the twentieth century, see William G. Rothstein, *American Physicians in the Nineteenth Century: From Sects to Science* (Baltimore, 1972), 310, and passim.

43. For overviews of twentieth-century medical development, see Paul Starr, *The Social Transformation of American Medicine* (New York, 1982); and Rosemary Stevens, *American Medicine and the Public Interest* (New Haven, 1971). On the somewhat paradoxical and poorly thought-out efforts of the federal government and the federal courts to undo during the 1970s and 1980s some of the long-term implications of the *Dent* decision, see Carl F. Ameringer, *The Health Care Revolution: From Medical Monopoly to Market Competition* (Berkeley, CA, 2008).

Epilogue

1. *Wheeling Register* (Wheeling, WV), March 8 and 9, 1891; *New York Times*, March 8, 1891.

2. *Wheeling Register* (Wheeling, WV), March 8 and 9, 1891.

3. *Wheeling Register* (Wheeling, WV), March 12, 1891.

4. *Wheeling Register* (Wheeling, WV), March 10, 1891; May 1, 1891.

5. *Wheeling Register* (Wheeling, WV), May 17, 1891.

6. *New York Times*, May 10, 1891.

7. *Wheeling Register* (Wheeling, WV), May 24, 1891.

8. *Wheeling Register* (Wheeling, WV), May 17, 1891.

9. *Wheeling Register* (Wheeling, WV), May 28, 1891; *New York Times*, May 17, 1891.

10. *Intelligencer* (Wheeling, WV), April 30, and May 2–9, 1892.

11. Circuit Court of Ohio County, West Virginia, "George Garrison for Murder," ledgers for May 27 and 28, 1892; *Intelligencer* (Wheeling, WV), May 17–30, 1892.

abortion, 63–64
Allen, Benjamin, 58, 128
American Medical Association (AMA), 16–18, 20, 84, 90–91, 141, 156–57; campaign for license laws before *Dent*, 21, 63–64, 67, 76, 103, 123, 125, 127–30; campaign for license laws after *Dent*, 161–64, 177–78; Code of Ethics, 31, 34–35, 38, 51–52, 130, 146; MSWV commitment to, 36–38, 40–47, 96, 103, 128; principles, endorsed before U.S. Supreme Court, 146–47, 177
American Medical College, St. Louis, 137
American Medical Eclectic College (AMEC), 100–102, 105, 137
American Public Health Association, 32, 45, 65, 103, 129
American Social Science Association, 166–67

Baird, George, 70, 78–79, 83, 156–58, 177; break with Garrison, 132–35; political leadership in MSWV, 46–48, 51–55, 58–60, 85, 129–30; shooting of, by Garrison, 3–5, 179–84
Baltimore and Ohio (B&O), 25, 36, 54, 69, 71; strike against, 45, 49; support for MSWV, 33, 41, 50, 64
Baltimore Medical College, 137
Barbee, Andrew, 60, 72, 74–75, 80, 85, 89
Bates, William, 34–35
Bee, Isaiah, 60, 78, 85
boards of health, 64–67. *See also* West Virginia Board of Health
Botanics, 13–14, 33–34, 49
Butchers' Union case, 149

Campbell, Matthew, 43–44, 55, 71
Carpenter, George, 98–99, 102
Chesapeake and Ohio (C&O), 25, 60, 69, 71
Columbus Medical College, 94–96, 135
constitutional issues, U.S.: bills of attainder, 118–19; Fourteenth Amendment, 105, 110, 116–17, 124, 141–49, 155, 165–70; police powers, 119–23, 144–49, 151, 165–69; Tenth Amendment, 9, 144
constitutional issues, West Virginia, 112–13, 119
Cooley, Thomas: cited in legal arguments, 121, 146; treatment of *Dent*, 167–68, 172
Crumrine, John, 60, 77
Culbertson, Hugh Emmett, 177
Cummins, Robert, 42
Cummins v. Missouri, 116, 141, 145, 150, 166–67

Dawson, William, 74, 198n26
Democratic Party, 104, 111; coming to power in WV, 44–45; divisions in, 44, 70, 78, 85, 131
Dent, Arthur Melville, 97, 99, 102–3, 130, 135–37; personal background, 52–53; rejection of license applications, 93–96; "Relative Merits" paper, 52–54, 58
Dent, George Washington, 52–53
Dent, Felix, 53
Dent, Frank Mortimer, 110, 125–26, 129, 136–38, 161, 163–64, 172; bringing suit against WV Board of Health, 103–5; denying guilt in public letter, 104; disputing with WV Board of Health, 98–102; personal

Dent, Frank Mortimer *(cont.)*
background, 53, 96–97; rejection of
license application, 102–3; and trial,
Preston County, 105–6; and trial,
U.S. Supreme Court, 124, 141–52; and
trial, WV Supreme Court of Appeals,
109, 112–23
Dent, Marmaduke Herbert, 52, 168;
arguments before U.S. Supreme
Court, 141–43; arguments before WV
Supreme Court of Appeals, 109–19;
attacking Reeves, 109, 115–16;
personal background, 103–4; and
petition to physicians, 109–11; and
Preston County circuit court,
104–6
Dent, William Marmaduke, 52–53, 68,
93, 157; and final break with Reeves,
99; offering alternative approach to
science, 54–57; as president of MSWV,
54–55
Dent family, 52–53
Dent v. West Virginia, 2, 124; reaction
of legal scholars to decision, 165–69,
172; reaction of physicians to
decision, 161–65; trial in U.S.
Supreme Court, 141–52; trial in WV
Supreme Court of Appeals, 112–23

Eclectics, 16, 19, 27, 49, 76, 100–2, 110,
147; colleges in Cincinnati, 100–101;
medical practices of, 13–14; reactions
to *Dent* decision, 163–65, 178
examinations, for medical license,
67–68, 82–83, 93, 114–15
Ex parte Garland, 117–18, 141, 167

Ferguson, James, 60, 63; actions in
legislature, 75, 77–78, 81, 85–89;
alliance with Reeves, 69–72, 156;
personal background, 69
Field, Stephen, 139, 141, 152, 161;
confusion of legal scholars by,
165–69; opinion in *Dent* case,
147–52
Fortney, Neil, 104
Freund, Ernst, 168–69
Frissell, John, 48, 52
Fuller, Horace, 169
Fuller, Melville, 139

Garrison, George, 83, 93, 130; appointed to state board, 132–34;
dispute over degree, 135–37; shooting
of Baird by, 3–5; trial of, 179–84
Green, Thomas, 111, 124–25, 139;
opinion in *Dent* case, 120–23

Hall, Silas, 42–44
Hildreth, Eugenius Augustus, 39,
49–50
Hoadly, George, 166–67
Homoeopaths, 19, 40, 76; medical
practices of, 13–14; reaction to *Dent*
decision, 163–65
Hopkins, Seymore, 83, 93–94, 96
Hupp, John, disputing Reeves, 33–34,
38, 50–52, 54, 58
Hydropaths, 13, 27, 44

Irons, B. F., 60
Irons, John, 174–75

Jackson, Jacob, 78, 80, 85, 112, 126,
131
Jacob, John, 44–45
Jarvis, Ann Maria Reeves, 29
Jefferson Medical College, 4, 55, 58,
128; and dispute over Garrison's MD
degree, 134–37
Johnson, Okey, 111

Kanawha Ring, 69, 72, 131
Koch, Robert, 157, 159

Lanham, Thomas, 101–2
Lazzelle, James, 39–42
licenses, medical: criteria for obtaining,
in 1882 WV, 67–68; early American,
15–16; following *Dent* decision,
155–65, 177–78; implications of
preparation-only, 172–77, 210n33;
numbers of, issued during 1880s, 83,
126–27, 137; revocations, 175–76

malpractice, medical, 42, 176–77
Matthews, Henry, 45
Medical Society of West Virginia
(MSWV): attitudes toward the
public, 39–43, 57, 75, 156–57; Board
of Censors, 35, 38, 46, 51–52, 58, 94,

130; doctrinal rigidity, 33–35, 38–39, 59, 128; founding of, 28, 32–34; legislative activity, 44–48, 59–60, 74; number of members, 37–38, 49; sectional tensions in, 36–37, 49–51, 54, 58–59. *See also* Baird, George; Reeves, James
Meyers, W. J., 168
McGrew, William, 73–75
Moffett, George, 83, 93, 95–96, 127
Mother's Day, 29
Mott, Rodney, 172
MSWV. *See* Medical Society of West Virginia

ordinary occupations, as legal concept, 148, 150, 158, 170, 209n29

Pasteur, Louis, 157, 159
physician-delegates, in state legislature, 60, 76–79, 85–87
police powers. *See* constitutional issues, U.S.
preparation-only licenses, implications of, 172–77, 210n33
professions: distinction between cultural and legal definitions of, 2, 16–18, 42, 169–71, 191n13; legal relationship to ordinary occupations, 148, 150, 158, 170, 209n29

railroads, ties to elite physicians, 70–71, 74. *See also* Baltimore and Ohio; Chesapeake and Ohio
Reeves, James: and Arthur Dent's application for medical license, 93–96; avowing licensing goal, 125–26; directing passage of Board of Health bill, 64–68, 72–75, 78–79, 90–91; disputing Hupp, 33–34, 38, 50–52, 54, 58; doctrinal fanaticism, 33, 38, 58, 80; and Frank Dent's application for medical license, 96–103; personal background, 28–34; reacting to state supreme court decision, 123; regaining control of MSWV, 58–60; response to Marmaduke Dent's petition, 111; retirement, 129–30; and science, 30–32, 157–61

Regulars: attitudes toward science, 12–13, 16–17, 39, 157–61; and consolidation of American medicine after *Dent,* 164–65, 177–78; medical practices of, 11–13; reaction to *Dent* decision, 161–62. *See also* American Medical Association; Medical Society of West Virginia; Reeves, James
Republican Party, 44, 46, 75, 85, 89, 104, 113, 181
Richardson, C. T., 98–99, 102
Russell, Isaac, 169

science: attitude of Regulars toward, 12–13, 16–17, 39, 43, 157–61; dispute over approaches to, within MSWV, 54–55; role of, in passage of medical license laws, 157–61
Slaughter-House Cases, 145, 147–48
smallpox, 9, 156, 180; threat of, used by Reeves, 66, 81, 88–89, 126
Snyder, Adam, 111
Starling Medical College, 53–54, 93–94, 108
State v. Ragland (WV), 138
Steere, David, 60, 86
Supreme Court, U.S., 1–2, 21, 48, 68, 79, 96, 105, 116, 123–24, 138; *Butchers' Union,* 149; *Cummins v. Missouri,* 116–17; *Dent v. West Virginia,* 2, 124, 139–52; *Ex parte Garland,* 117; justices serving on, in 1888, 139–40; *Slaughter-House Cases,* 145, 147–48

ten-year clause: debated in court, 110, 114, 116, 118; Frank Dent denied under, 98–102; in license law, 68, 77, 90, 126–27, 136–37
Thomsonians, 13, 49

Ulrich, Charles, 75, 171, 182–83
U.S. Supreme Court. *See* Supreme Court, U.S.

Van Kirk, William, 55, 58–59, 71

Watts, Cornelius, 119–21
West Virginia Board of Health: criminal sanctions of, 68–69; criteria for licenses, 67–68; dispute with

West Virginia Board of Health *(cont.)*
governor over Garrison appointment,
130–37; establishing guidelines for
reputability of medical colleges,
95–96; initial appointments to, 80;
legal powers of, 64–69; MSWV
endorsing creation of, 45; qualifications of members, 66–67, 73, 75–76;
soliciting voluntary contributions, 82
West Virginia Supreme Court of
Appeals, 104, 106, 109, 111; decision
of, in *Dent* case, 120–23, 125,

129–30; decision of, in *State v.
Ragland,* 138; justices on, 111–12
Wheeling: city's physicians protesting
Garrison appointment, 133–34;
number of physicians in, 1870, 27
Wilson, Emanuel: as governor, 131–35,
137, 180; in legislature, 86–87
Wilson, Louis, 79
Woods, Joseph, 72–74
Woods, Samuel, 111

Yick Wo v. Hopkins, 149, 151